Springer Series in Information Sciences 13

Editor: King-sun Fu

Springer Series in Information Sciences
Editors: King-sun Fu   Thomas S. Huang   Manfred R. Schroeder

Volume 1	**Content-Addressable Memories** By T. Kohonen	
Volume 2	**Fast Fourier Transform and Convolution Algorithms** By H. J. Nussbaumer 2nd Edition	
Volume 3	**Pitch Determination of Speech Signals** Algorithms and Devices By W. Hess	
Volume 4	**Pattern Analysis** By H. Niemann	
Volume 5	**Image Sequence Analysis** Editor: T. S. Huang	
Volume 6	**Picture Engineering** Editors: King-sun Fu and T. L. Kunii	
Volume 7	**Number Theory in Science and Communication** With Applications in Cryptography, Physics, Biology, Digital Information, and Computing By M. R. Schroeder	
Volume 8	**Self-Organization and Associative Memory** By T. Kohonen	
Volume 9	**An Introduction to Digital Picture Processing** By L. P. Yaroslavsky	
Volume 10	**Probability, Statistical Optics, and Data Testing** A Problem Solving Approach. By B. R. Frieden	
Volume 11	**Physical and Biological Processing of Images** Editors: O. J. Braddick and A. C. Sleigh	
Volume 12	**Multiresolution Image Processing and Analysis** Editor: A. Rosenfeld	
Volume 13	**VLSI for Pattern Recognition and Image Processing** Editor: King-sun Fu	
Volume 14	**Mathematics of Kalman-Bucy Filtering** By P. A. Ruymgaart and T. T. Soong	

VLSI for Pattern Recognition and Image Processing

Editor: King-sun Fu

With 114 Figures

Springer-Verlag
Berlin Heidelberg New York Tokyo 1984

Professor King-sun Fu
School of Electrical Engineering, Purdue University,
West Lafayette, IN 47907, USA

Series Editors:

Professor King-sun Fu
School of Electrical Engineering, Purdue University,
West Lafayette, IN 47907, USA

Professor Thomas S. Huang
Department of Electrical Engineering and Coordinated Science Laboratory,
University of Illinois, Urbana, IL 61801, USA

Professor Dr. Manfred R. Schroeder
Drittes Physikalisches Institut, Universität Göttingen, Bürgerstraße 42–44,
D-3400 Göttingen, Fed. Rep. of Germany

ISBN 3-540-13268-6 Springer-Verlag Berlin Heidelberg New York Tokyo
ISBN 0-387-13268-6 Springer-Verlag New York Heidelberg Berlin Tokyo

Library of Congress Cataloging in Publication Data. Main entry under title: VLSI for pattern recognition and image processing. (Springer series in information sciences ; 13) 1. Image processing–Digital techniques. 2. Pattern recognition systems. 3. Integrated circuits–Very large scale integration. I. Fu, K.S. (King Sun), 1930-. II. Agrawal, D.P. III. Title: V.L.S.I. for pattern recognition and image processing. IV. Series: Springer series in information sciences ; v. 13. TA1632.V57 1984   621.36'7   84-3910

This work is subject to copyright. All rights are reserved, whether the whole or part of the material is concerned, specifically those of translation, reprinting, reuse of illustrations, broadcasting, reproduction by photocopying machine or similar means, and storage in data banks. Under § 54 of the German Copyright Law where copies are made for other than private use, a fee is payable to "Verwertungsgesellschaft Wort", Munich.

© by Springer-Verlag Berlin Heidelberg 1984
Printed in Germany

The use of registered names, trademarks, etc. in this publication does not imply, even in the absence of a specific statement, that such names are exempt from the relevant protective laws and regulations and therefore free for general use.

Offset printing: Beltz Offsetdruck, 6944 Hemsbach/Bergstr. Bookbinding: J. Schäffer OHG, 6718 Grünstadt
2153/3130-543210

Preface

During the past two decades there has been a considerable growth in interest in problems of pattern recognition and image processing (PRIP). This interest has created an increasing need for methods and techniques for the design of PRIP systems. PRIP involves analysis, classification and interpretation of data. Practical applications of PRIP include character recognition, remote sensing, analysis of medical signals and images, fingerprint and face identification, target recognition and speech understanding.

One difficulty in making PRIP systems practically feasible, and hence, more popularly used, is the requirement of computer time and storage. This situation is particularly serious when the patterns to be analyzed are quite complex. Thus it is of the utmost importance to investigate special computer architectures and their implementations for PRIP. Since the advent of VLSI technology, it is possible to put thousands of components on one chip. This reduces the cost of processors and increases the processing speed. VLSI algorithms and their implementations have been recently developed for PRIP. This book is intended to document the recent major progress in VLSI system design for PRIP applications.

West Lafayette, March 1984   *King-sun Fu*

Contents

1. Introduction. By K.S. Fu (With 1 Figure) 1
   1.1 VLSI System 1
   1.2 VLSI Algorithms 3
   1.3 Summary of Book 3
   References 4

Part I General VLSI Design Considerations

2. One-Dimensional Systolic Arrays for Multidimensional Convolution and Resampling. By H.T. Kung and R.L. Picard (With 11 Figures) .. 9
   2.1 Background 9
   2.2 Systolic Convolution Arrays 11
   2.3 Variants in the Convolution Problem 16
   2.4 Implementation 17
   2.5 Concluding Remarks 21
   References 24

3. VLSI Arrays for Pattern Recognition and Image Processing: I/O Bandwidth Considerations. By T.Y. Young and P.S. Liu (With 11 Figures) 25
   3.1 Background 25
   3.2 Arrays for Matrix Operations 27
   3.3 Arrays for Pattern Analysis 32
   3.4 Image-Processing Array 37
   3.5 Conclusions 40
   References 41

Part II VLSI Systems for Pattern Recognition

4. **VLSI Arrays for Minimum-Distance Classifications**
   By H.-H. Liu and K.S. Fu (With 10 Figures) 45
   4.1 Minimum-Distance Classification 45
   4.2 Vector Distances 48
   4.3 String Distances 53
   4.4 Examples of Application 60
   4.5 Summary 61
   References 62

5. **Design of a Pattern Cluster Using Two-Level Pipelined Systolic Array.** By L.M. Ni and A.K. Jain (With 12 Figures) 65
   5.1 Background 65
   5.2 Description of Squared-Error Pattern Clustering 67
   5.3 The Systolic Pattern Clustering Array 70
   5.4 System Operating Characteristics 78
   5.5 Conclusion 82
   References 83

6. **VLSI Arrays for Syntactic Pattern Recognition**
   By Y.P. Chiang and K.S. Fu (With 9 Figures) 85
   6.1 Pattern Description and Recognition 85
   6.2 Syntactic Pattern Recognition 87
   6.3 VLSI Implementation 94
   6.4 Simulation and Examples 99
   6.5 Concluding Remarks 102
   References 104

Part III VLSI Systems for Image Processing

7. **Concurrent Systems for Image Analysis.** By G.R. Nudd
   (With 15 Figures) 107
   7.1 Background 107
   7.2 Processing Requirements for Image Analysis 110
   7.3 Concurrent VLSI Architectures 113
   7.4 Conclusions 131
   References 131

8. *VLSI Wavefront Arrays for Image Processing*
   By S.-Y. Kung and J.T. Johl (With 8 Figures) 133
   8.1 Background 133
   8.2 Parallel Computers for Image Processing 136
   8.3 Design of Pipelined Array Processors 139
   8.4 Image-Processing Applications 143
   8.5 Wafer-Scale Integrated System 153
   8.6 Conclusion 154
   References 154

9. *Curve Detection in VLSI.* By M.J. Clarke and C.R. Dyer
   (With 4 Figures) 157
   9.1 Background 157
   9.2 Montanari's Algorithm 158
   9.3 Systolic Architectures 159
   9.4 A VLSI Design of the Column Pipeline Chip 168
   9.5 SIMD Array Algorithm 171
   9.6 Concluding Remarks 172
   References 173

10. *VLSI Implementation of Cellular Logic Processors*
    By K. Preston, Jr., A.G. Kaufman, and C. Sobey (With 14 Figures) . 175
    10.1 Cellular Logic Processors 175
    10.2 Binary Neighborhood Functions 176
    10.3 CELLSCAN 178
    10.4 The Golay Parallel Pattern Transform 180
    10.5 The diff3-GLOPR 182
    10.6 The Preston-Herron Processor (PHP) 183
    10.7 The LSI/PHP 186
    References 193

11. *Design of VLSI Based Multicomputer Architecture for Dynamic Scene
    Analysis.* By D.P. Agrawal and G.C. Pathak (With 4 Figures) 195
    11.1 Background 195
    11.2 Dynamic Scene Analysis Algorithm 196
    11.3 Existing Multicomputer Architecture 200
    11.4 Scheduling and Parameters of Interest 205
    11.5 Performance Evaluation 206
    11.6 Conclusion 207
    References 207

12. **VLSI-Based Image Resampling for Electronic Publishing**
    By Z.Z. Stroll and S.-C. Kang (With 15 Figures) 209
    12.1 Introduction to the "Electronic Darkroom" 209
    12.2 Overview of System Design Concepts 210
    12.3 Resampling Algorithms 213
    12.4 System Architecture and Performance 220
    12.5 Trade-Offs and Conclusions 228
    References 229

Subject Index 231

List of Contributors

Agrawal, Dharma P.
  Department of Electrical and Computer Engineering, North Carolina State University, Raleigh, NC 27695, USA

Chiang, Yetung P.
  Department of Electrical & Computer Engineering, Washington State University, Pullman, WA 99164, USA

Clarke, Martin J.
  Martin Marietta Aerospace, Box 5837, Orlando, FL 32855, USA

Dyer, Charles R.
  Computer Sciences Department, University of Wisconsin Madison, WI 53706, USA

Fu, King-sun
  School of Electrical Engineering, Purdue University West Lafayette, IN 47907, USA

Jain, Anil K.
  Department of Computer Science, Michigan State University East Lansing, MI 48824, USA

Johl, John T.
  Department of Electrical Engineering Systems, University of Southern California, Los Angeles, CA 90089, USA

Kang, Sang-Chul
  Electronics Division, Xerox Corp., El Segundo, CA 90245, USA

Kaufman, Amy G.
  Department of Electrical Engineering, Carnegie-Mellon University Pittsburgh, PA 15213, USA

Kung, Hsiang T.
  Department of Computer Science, Carnegie-Mellon University Pittsburgh, PA 15213, USA

Kung, Sun-Yuan
  Department of Electrical Engineering Systems, University of Southern California, Los Angeles, CA 90089, USA

Liu, Hsi-Ho
   Department of Electrical and Computer Engineering, University of Miami,
   Coral Gables, FL 33124, USA

Liu, Philip S.
   Department of Electrical and Computer Engineering, University of Miami
   Coral Gables, FL 33124, USA

Ni, Lionel M.
   Department of Computer Science, Michigan State University
   East Lansing, MI 48824, USA

Nudd, Graham R.
   Hughes Research Labs, 3011 Malibu Canyon Rd., Malibu, CA 90265, USA

Pathak, Girish C.
   Department of Electrical and Computer Engineering, North Carolina
   State University, Raleigh, NC 27695, USA

Picard, Ray L.
   ESL Incorporated, A Subsidiary of TRW Inc., 495 Java Drive, P.O. Box 3510
   Sunnyvale, CA 94088-3510, USA

Preston, Kendall, Jr.
   Department of Electrical Engineering, Carnegie-Mellon University
   Pittsburgh, PA 15213, USA

Sobey, Charles
   Department of Electrical Engineering, Carnegie-Mellon University
   Pittsburgh, PA 15213, USA

Stroll, Zoltan Z.
   Defense Systems Group of TRW Inc., One Space Park, 02/2791
   Redondo Beach, CA 90278, USA

Young, Tzay Y.
   Department of Electrical and Computer Engineering, University of Miami
   Coral Gables, FL 33124, USA

1. Introduction

K.S. Fu

1.1 VLSI System

The Very Large Scale Integration (VLSI) technology [1.1] enables many hundreds of thousands of switching devices to be placed on a single silicon chip with feature sizes approaching 1 micrometer. This advancement of IC technology promises the usage of much-smaller-sized transistors and wires, and has also increased the density of the circuitry. In addition, the reduced size results in faster and/or lower power operations [1.2,3].

Some commercial examples are: Motorola's MC68000 or the 64 K memory chip, HP's 32-bit CMOS processor [1.4] and Intel's iAPX-432 processor [1.5]. As for research projects, RISC [1.6] and MIPS [1.7] are general purpose processors designed to be implemented on a single VLSI chip. These two projects emphasize processors with reduced and simple-instruction sets, they can achieve high throughput. Note that all the systems mentioned so far are sequential computers. The X-tree system [1.8], on the other hand, is a MIMD type parallel computer with each processor, the so called X-node, built upon a single VLSI chip. The systolic searching tree [1.9] is a binary tree-structure system designed for easy implementation with VLSI technology. This tree system is efficient for search problems and operates in a SIMD node.

Special VLSI chips are usually constructed only when they can solve a problem satisfying two criteria; the problem is currently very time-consuming, and the proposed special-purpose device is much more efficient than conventional ways of solving this problem. Therefore, almost all the specially designed VLSI chips take advantage of pipelining and parallelism. The most significant contribution in this area is the concept of a systolic array and many of its applications [1.10,11]. Figure 1.1 demonstrates six different VLSI architectures. The first three parts Fig.1.1a-c, are also called mesh-connected processor arrays. These arrays have simple and regular interconnections, which lead to cheap implementations and high densities. High density implies both high performance and a low overhead for support components. The numerous

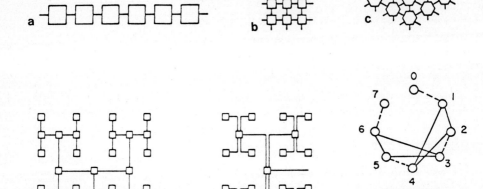

Fig.1.1a-f. VLSI system architectures. a) A one-dimensional linear array; b) a two-dimensional square array; c) a two-dimensional hexagonal array; d) a binary tree; e) a quad-tree; f) a shuffle-exchange network

applications of these arrays have been well discussed. The tree structures are depicted in Fig.1.1d-e. In a tree, the number of processors available increases exponentially at each level. If the problem to be solved has this growth pattern, then the tree geometry will be suitable for the problem. The binary tree system has an interesting aspect, that is, any particular processor in a tree of n processors can be accessed in at most $\log_2 n$ time. This is favorable when comparing with the $O(n)$ access time for a 1-D array or the $O(N^{\frac{1}{2}})$ access time for a rectangular array. The shuffle-exchange network shown in Fig.1.1f is also mesh-connected. It should be mentioned that the particular communication structure of this network makes the system suitable for bitonic sort and n-point Fast Fourier Transform (FFT) [1.12]. However, the shuffle-exchange network, although it has great power in permutation, suffers from a very low degree of regularity and modularity. This can be a serious drawback for VLSI implementation. Consequently, mesh-connected arrays and trees are favorable architectures for special purpose VLSI design.

Many specially designed VLSI arrays and trees are proposed for pattern recognition and image processing tasks. Recently, a few VLSI processors were proposed especially for image processing.

1.2 VLSI Algorithms

Since the advent of VLSI technology, it is possible to put thousands of gates on one chip. This reduces the cost for processors and increases the communication speed. It also changes the criteria for algorithm design [1.13,14]. A good VLSI algorithm has to meet the following requirements: 1) the function of each processor must be kept simple and perform constant-time operation, 2) the communication geometry should be simple and regular, and 3) the data movement should be kept simple, regular and uniform. As mentioned in Sect. 1.1, the mesh-connected arrays and trees are favorable architectures for special purpose VLSI design. Besides the tree and the 1-D array, all other arrays have the mixed blessing of pipelining and parallelism. In order to take full advantage of this computational capability, VLSI algorithms are usually proposed with specific configuration in mind.

In their book *Mead* and *Conway* [Ref.1.1, Sect.8.3] explicitly introduce the VLSI algorithms and their associated processor arrays. For instance, the matrix-vector multiplication on linear arrays; the matrix multiplication, and LU-decomposition on hexagonal arrays; the color-cost problem on tree-structure systems, etc. *Kung* [1.13] provided a table which outlines the wide applications of VLSI algorithms. Many examples were given in his paper like an odd-even transposition sort on linear arrays; search algorithms on a tree machine (also see [1.15]).

Besides the applications mentioned above, VLSI algorithms have seen various applications in pattern recognition and image processing. For example, the pattern matching problem on a 2-D array and multi-dimensional convolution on a two-level linear array; FIR (finite impulse response) filtering and DFT (discrete Fourier transform) [1.16] on a linear array; recognition of finite-state languages on an 1-D array; the running-order statistics problems (a generalization of median smoothing) on a linear array; dynamic programming for optimal parenthesization and CYK (Cocke-Younger-Kasami) parsing on a 2-D triangular array; the convolution of two finite sequences on a tree system, etc.

1.3 Summary of Book

Many pattern-recognition and image-processing algorithms have been regarded computationally expensive. Consequently their utility in real-time applications is often restricted. With recent advances in VLSI technology, design and implementation of VLSI systems for pattern recognition and image processing have received increasing attention. This book is intended to document the recent major progress in VLSI system design for pattern recognition and image processing.

The contents of the book can be roughly divided into three parts. The first part deals with general VLSI design considerations. H.T. Kung presents some experiences in the implementing of highly parallel processors using VLSI technology. T.Y. Young and P.S. Liu discuss the I/O bandwidth considerations in VLSI arrays for pattern recognition and image processing.

The second part contains three chapters that are devoted to the VLSI system design for pattern recognition. H.H. Liu and K.S. Fu describe the design of VLSI arrays for minimum vector-distance and string-distance classifications. L.M. Ni and A.K. Jain present a two-level pipelined systolic array for pattern cluster analysis. VLSI arrays for syntactic pattern recognition algorithms are treated by Y.T. Chiang and K.S. Fu.

The third part consists of six chapters that are concerned with the use of VLSI systems for image processing. Concurrent systems for image analysis are discussed by G.R. Nudd. S.Y. Kung presents the results of using VLSI wave-front arrays for image processing. C.R. Dyer and M.J. Clarke propose four VLSI designs for line and curve detection in images. A VLSI implementation of cellular logic processors is described by K. Preston. Dynamic scene analysis using a VLSI-based multicomputer architecture is treated by D.P. Agrawal and G.C. Pathak. Z. Stroll and S.C. Kang propose a VLSI system of image resampling for electronic publishing.

References

1.1 C.V. Ramamoorthy, Y.W. Ma: Impact of VLSI on Computer Architectures, in *VLSI Electronics 3*, ed. by N.G. Einsprunch (Academic, New York 1982) pp.2-22
1.2 R. Egan: The Effect of VLSI on Computer Architecture. Computer Architecture News **10**, 19-22 (1982)
1.3 C.A. Mead, L.A. Conway: *Introduction to VLSI Systems* (Addison-Wesley, Reading, MA 1980)
1.4 J.W. Beyers et al.: "A 34 Bit VLSI CPU Chip", Digest of Papers. IEEE Int'l Solid State Circutis Conf. (Feb. 1981) pp.104-105
1.5 J. Rattner, W.W. Lattin: Ada Determines Architecture of 32-bit Microprocessor. Electronics **54**, 119-126 (1981)
1.6 D.T. Fitzpatrick et al.: "VLSI Implementations of a Reduced Instruction Set Computer", in *VLSI Systems and Computations*, ed. by H.T. Kung, B. Sproull and G. Steele (Computer Science Press 1981)
1.7 J. Hennessy et al.: MIPS: A VLSI Processor Architecture, in *VLSI Systems and Computations*, ed. by H.T. Kung, B. Sproull and G. Steele (Computer Science Press 1981)
1.8 A.M. Despain: X-Tree: A Multiple Microcomputer System, Proc. COMPCON SPRING (Feb. 1980) pp.324-327

1.9 J.L. Bentley, H.T. Kung: A Tree Machine for Searching Problems, Proc. 1979 Int'l Conf. Parallel Processing (Aug. 1979) pp.257-266
1.10 H.T. Kung, C.E. Leiserson: "Algorithms for VLSI Processor Arrays", in *Introduction to VLSI Systems*, ed. by C. Mead and L. Conway (Addison-Wesley, Reading, MA 1980) Sect. 8.3, pp.271-291
1.11 M.J. Foster, H.T. Kung: "The Design of Special-Purpose VLSI Chips", Computer **13**, 26-40 (Jan. 1980)
1.12 H.J. Nussbauer: *Fast Fourier Transform and Convolution Algorithms*, 2nd ed., Springer Ser. Inform. Sci., Vol.2 (Springer, Berlin, Heidelberg, New York 1982)
1.13 H.T. Kung: The Structure of Parallel Algorithms, in *Advances in Computers*, Vol.19, ed. by M.C. Yovits (Academic, New York 1980)
1.14 H.T. Kung: Let's Design Algorithms for VLSI Systems, Proc. Caltech Conf. on VLSI (Jan. 1979) pp.65-90
1.15 C.E. Leiserson: Systolic Priority Queues, Proc. Caltech Conf. on VLSI (Jan. 1979)
1.16 H.J. Nussbauer: In *Two-Dimensional Digital Signal Processing II*, ed. by T.S. Huang. Topics Appl. Phys., Vol.43 (Springer, Berlin, Heidelberg, New York 1981)

Part I

General VLSI Design Considerations

2. One-Dimensional Systolic Arrays for Multidimensional Convolution and Resampling

H.T. Kung[1] and R.L. Picard

We present one-dimensional systolic arrays for performing two- or higher-dimensional convolution and resampling. These one-dimensional arrays are characterized by the fact that their I/O bandwidth requirement is independent of the size of the convolution kernel. This contrasts with alternate two-dimensional array solutions, for which the I/O bandwidth must increase as the kernel size increases. The proposed architecture is ideal for VLSI implementation—an arbitrarily large kernel can be handled by simply extending the linear systolic array with simple processors of the same type, so that one processor corresponds to each kernel element.

2.1 Background

Multidimensional convolution and the related resampling computation constitute some of the most compute-intensive tasks in image processing. For example, a two-dimensional (2-D) convolution with a general 4×4 kernel would require 16 multiplications and 15 additions to be performed for generating each output pixel. To perform this convolution on a 1000×1000 image at the video rate would require the computing power of over 100 VAX-11/780s. If the kernel is larger or the dimensionality higher, even more computation power would be required. Though computationally demanding, multidimensional convolution and resampling are highly regular computations. We will show how to exploit this regularity and build cost-effective, high-throughput pipelined systems to perform 2-D and higher-dimensional convolution and resampling.

These pipelined designs are based on the systolic array architecture [2.1,2]. A systolic array organizes a regular computation such as the 2-D convolution through a lattice of identical function modules called cells. Unlike other parallel processors employing a lattice of function modules,

---
1 H.T. Kung was supported in part by the Office of Naval Research under Contracts N00014-76-C-0370, NR 044-422 and N00014-80-C-0236, NR 048-659

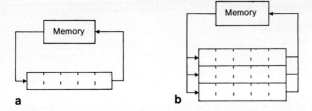

Fig.2.1. (a) 1-D systolic array, and (b) 2-D systolic array

A systolic array is characterized by a regular data flow. Typically, two or more data streams flow through the cells of the systolic array in various speeds and directions. Data items from different streams interact with each other at the cells where they meet. From the user's point of view, a computation can be performed by simply pumping the input data streams into one end of the systolic array, and then collecting results from the other end. The crux of the systolic array approach is to ensure that once a data item is brought out from the system memory it can be used effectively at each processing cell it passes. Therefore a high computation throughput can be achieved with modest memory bandwidth. Being able to use each input data item a number of times is just one of the many advantages of a systolic architecture. Other advantages, including modular expandability, simple and regular data and control flows, use of simple and uniform cells, and elimination of global broadcasting and fan-in, are also characteristic [2.1]. These properties are highly desirable for VLSI implementations; indeed the advances in VLSI technology have been a major motivation for systolic architecture research.

A systolic array can be a one-dimensional (1-D) or two-dimensional (2-D) array of cells, as depicted in Fig.2.1. We see that for the 2-D structure, more than one cell inputs from or outputs to the system memory simultaneously. Thus, if the I/O speed is higher than the cell speed, the 2-D structure could be used. *Kung* and *Song* [2.3] describe a prototype chip that implements a systolic array for the 2-D convolution. In their design, because cells are implemented bit-serial, they are relatively slow; as a result the 2-D systolic structure is used.

In the present paper, we address a different scenario where the communication bandwidth between the system memory and the systolic array is a limiting factor. This situation typically arises when cells are implemented by high-speed bit-parallel logic and it is infeasible or too expensive to include a sufficiently large buffer memory and its control on the same chips or subsystems that host the systolic array, to reduce the I/O bandwidth requirement. In this case the bandwidth of memory, bus or pin, rather than the speed of the cells, is the bottleneck for achieving high system throughput. Therefore, it is appropriate to use the 1-D systolic structure, that has the minimum possible I/O bandwidth requirement. In Sect.2.2, we present the fundamental result of this paper—fully utilized 1-D systolic arrays for performing 2-D and higher-dimensional convolutions. Extensions of this result to other related problems, such as resampling, are given in Sects.2.3,4, where the requirements imposed on systolic array implementations by various classes of convolutions are also discussed. In Sect.2.5 are concluding remarks and a list of problems that have systolic solutions.

2.2 Systolic Convolution Arrays

We first review a 1-D systolic array for the 1-D convolution, and then show how this basic design can be extended to a 1-D systolic array for 2-D, 3-D and higher-dimensional convolutions.

2.2.1 Systolic Array for 1-D Convolution

The 1-D convolution problem is defined as follows:

Given the kernel as a sequence of weights $(w_1, w_2, \ldots, w_k)$ and the input sequence $(x_1, x_2, \ldots, x_n)$,
compute the output sequence $(y_1, y_2, \ldots, y_{n+1-k})$, defined by $y_i = w_1 x_i + w_2 x_{i+1} + \ldots + w_k x_{i+k-1}$.

For simplicity in illustration, we assume that the kernel size k is three. It should be evident that our results generalize to any positive integer k. Depicted in Fig.2.2 is one of the many possible 1-D systolic arrays for the 1-D convolution [2.1]. Weights are preloaded into the array, one for each cell. Both partial results y_i and inputs x_i flow from left to right, but the y_i move twice as fast as the x_i, so that each y_i can meet three x_i. More precisely, each x_i stays inside every cell it passes for one cycle, thus it takes twice as long to march through the array as a y_i does. It is easy to see that each y_i, initialized to zero before entering the leftmost cell, is able to accumulate all its terms while marching to the right. For example, y_1 accumulates $w_3 x_3, w_2 x_2$, and $w_1 x_1$ in three consecutive cycles at the lestmost, middle, and rightmost cells, respectively.

This 1-D systolic array can be readily used to form a 2-D systolic array for the 2-D convolution by essentially stacking copies of it in the vertical direction [2.3], as depicted in Fig.2.3. However, this 2-D systolic structure

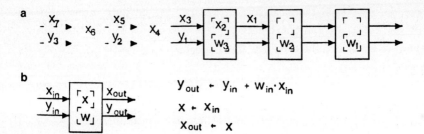

Fig.2.2. (a) Systolic array for 1-D convolution using a kernel of size 3, and (b) the cell definition

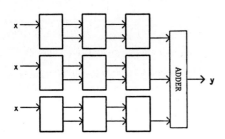

Fig.2.3. 2-D systolic array obtained by stacking 1-D arrays

has high I/O bandwidth requirements when the kernel size k is large, and thus is inappropriate for the scenario addressed in this paper. The task we face is to design a new 1-D systolic array that is capable of performing the 2-D convolution with full utilization of all its cells. The next section shows how this can be accomplished.

2.2.2 1-D Systolic Array for 2-D Convolution

The 2-D convolution problem is defined as follows:

Given the weights w_{ij} for $i,j = 1,2,\ldots,k$ that form a $k \times k$ kernel, and input image x_{ij} for $i,j = 1,2,\ldots,n$,
compute the output image y_{ij} for $i,j = 1,2,\ldots, n$ defined by

$$y_{ij} = \sum_{\ell=1}^{k} \sum_{h=1}^{k} w_{h,\ell} x_{i+h-1, j+\ell-1}$$

Assuming k = 3, it is easy to show that the 2-D convolution can be accomplished by performing the following three 1-D convolutions, all using $(w_{11}, w_{21}, w_{31}, w_{12}, w_{22}, w_{32}, w_{13}, w_{23}, w_{33})$ as the sequence of weights:

1) Computing $(y_{11}, y_{12}, y_{13}, y_{14}, \ldots)$ using $(x_{11}, x_{21}, x_{31}, x_{12}, x_{22}, x_{32}, x_{13}, x_{23}, x_{33}, x_{14}, x_{24}, x_{34}, \ldots)$ as the input sequence.

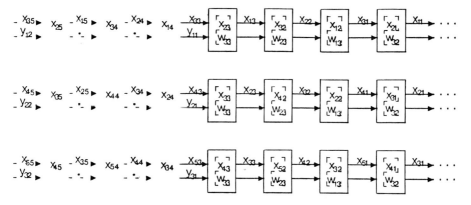

Fig.2.4. Three 9-cell 1-D systolic arrays

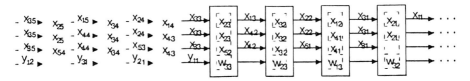

Fig.2.5. Systolic array obtained by combining the three systolic arrays of Fig.2.4

2) Computing $(y_{21}, y_{22}, y_{23}, y_{24}, \ldots)$ using $(x_{21}, x_{31}, x_{41}, x_{22}, x_{32}, x_{42}, x_{23}, x_{33}, x_{43}, x_{24}, x_{34}, x_{44}, \ldots)$ as the input sequence.

3) Computing $(y_{31}, y_{32}, y_{33}, y_{34}, \ldots)$ using $(x_{31}, x_{41}, x_{51}, x_{32}, x_{42}, x_{52}, x_{33}, x_{43}, x_{53}, x_{34}, x_{44}, x_{54}, \ldots)$ as the input sequence.

Using the result of the preceding section, each of these 1-D convolutions can be carried out by a 1-D systolic array consisting of 9 cells, as depicted in Fig.2.4.

Notice that a cell in any of these 1-D arrays is involved in computing a y_{ij} only one third of the time. This suggests that the three systolic arrays be combined into one. Figure 2.5 displays the combined systolic array having full utilization of all its cells. Each cell in this combined array is time-shared to perform the operations of the corresponding cells in the original systolic arrays. Observe that the three x data streams in the systolic array carry redundant information — two of the three x values entering a cell are always equal. Figure 2.6 shows the final 1-D systolic array for which only two x data streams are used, but each cell now has to choose one of the two available x input values for use at every cycle. By having the x_{ij} with odd j flow in the top x data stream and x_{ij} with even j flow in the bottom x data

13

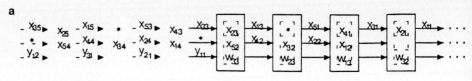

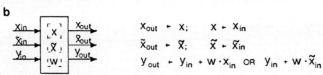

Fig.2.6. (a) 1-D systolic array for 2-D convolutions and (b) the cell definition

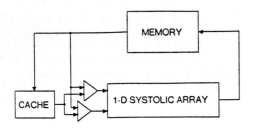

Fig.2.7. Use of cache to eliminate reaccess of input pixels

stream, as shown in the figure, the choice of the x input at each cell is easily determined. The simple rule is that a cell should use an x data stream for three consecutive cycles before switching to the other stream, and continue alternating in this manner.

For 2-D convolutions with general $k \times k$ kernels, the input image is fed to the systolic array in columns $2k-1$ pixels high, and the output image is generated by the systolic array in swaths which are k pixels high. In generating adjacent output swaths, some input pixels are fed into the systolic array twice. To avoid having to bring these pixels out from the system memory more than once, a cache that can hold $k-1$ lines of input pixels can be used[2], as shown in Fig.2.7.

In summary, to compute any 2-D convolution using a $k \times k$ kernel, a 1-D systolic array with k^2 cells and two input data streams can be constructed such that all the cells can be fully utilized during the computation. For the special case where the kernel forms a rank one matrix, the 2-D convolution can be carried out in two stages — a 1-D convolution along one dimension followed by another 1-D convolution along the other dimension. Therefore this

---

[2] By an information theoretic argument one can in fact show that any convolution device that inputs each pixel of a given image of width n only once must use some form of buffer memory of total size $O(kn)$ [2.4].

degenerate case requires only 1-D systolic arrays with k cells for each of the 1-D convolutions.

2.2.3 1-D Systolic Array for Multidimensional Convolution

The above 1-D systolic design for 2-D convolution can be generalized to handle any multidimensional convolution. For illustration, we consider here the three-dimensional (3-D) convolution. Denote by $w_{ij}^{(h)}$, $x_{ij}^{(h)}$, and $y_{ij}^{(h)}$ the weights, input pixels and output pixels, respectively. Figure 2.8 shows how pixels are indexed for a 3-D image.

Suppose that the size of the kernel is $k \times k \times k$. Then the 1-D systolic array for the 3-D convolution contains k^3 cells and employs four input data streams. The input image is fed into the systolic array in $(2k-1) \times (2k-1)$ planes using a lexicographical order, and the output image is generated by the systolic array in swaths of k pixels high by k pixels deep. Figure 2.9 depicts the systolic array for the case when $k = 3$. At each cycle, a cell chooses one of the values that are available from the four input data streams. The rule is that a cell operates on data from each input data stream for three (k in general) consecutive cycles, and then switches to a

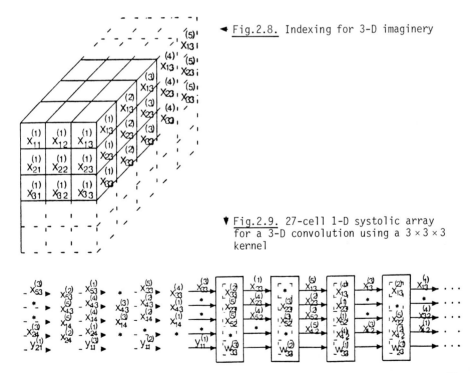

◀ Fig.2.8. Indexing for 3-D imaginery

▼ Fig.2.9. 27-cell 1-D systolic array for a 3-D convolution using a $3 \times 3 \times 3$ kernel

new stream. The order of choosing data streams is as follows: first, second, first; third, fourth, third; first, second, first; and so on. Therefore the switch between one pair of streams (the first and second) and the other (the third and fourth) occurs every nine (k^2 in general) cycles.

The above approach extends systematically to any number of dimensions. For example, for the four-dimensional convolution, the switch between one set of four input streams and the other occurs every k^3 cycles. This structured approach to multidimensional convolution can make an unwieldy problem solvable by providing an optimized hardware configuration, as well as optimized data access and control mechanisms.

2.3 Variants in the Convolution Problem

Signal-processing and image-processing applications make extensive use of 2-D convolution-like calculations for filtering, enhancement, interpolation, and other similar functions. Thus the 1-D systolic array for the 2-D convolution presented in the last section can potentially be used to implement any of these functions. Based on the dynamics of the convolutional kernels, these functions may be organized into three broad categories consisting of fixed, adaptive, and selectable kernels.

A fixed kernel convolution has only a single set of weights. Typically, this would be found in equipment custom-designed to perform one specific function, so the weights could be hard-wired into the machine. This could be the best alternative in detection or fixed-filtering applications where the primary objective is to minimize the recurring hardware cost.

The adaptive kernel class is a generalization of the fixed kernel class, the major difference being that the weights may be changed. This class divides further into two distinct subclasses. The first subclass is basically a fixed kernel, but where the weights may be initialized (down-loaded) with an arbitrary value whenever the convolver is quiescent. The second subclass, and one gaining a lot of attention in signal processing areas, is the "true" adaptive kernel. For this subclass, the weights may be altered by some external source without interrupting the on-going convolution process. In this case, the weigths would be updated to reflect a change in the convolver's environment, often based on the convolver's own input or result stream. The critical parameter for this type of kernel is the adaptation time constant, that is, the rate at which the weights may be modified. This adaptation rate can easily be more than an order of magnitude slower than the data rate, being limited almost entirely by the complexity of the weight calculation.

Typically, calculating adaptive weights requires time-consuming inversion of a time-varying covariance matrix [2.5], although other systolic arrays could be used to speed up this calculation [2.6]. Also important is the guarantee that no erroneous results are produced during the transition from one weight set to another.

The final classification is selectable kernels, distinguished from the adaptive kernels by allowing only a limited choice for the weights. Usually, the selectable kernels would be prestored in tables, instead of directly calculated, and the particular weights to be used are chosen by the value of a control variable associated with each desired output. Note that this permits the weights to change at the same rate as the data. This is a very important class of convolutions, including most forms of interpolation and resampling. This is also fundamental to the geometric correction or spatial warping of images.

In the next section, we discuss how the basic systolic design for the 2-D convolution, presented in Sect.2.2 can be reduced to practice for some of the applications mentioned above.

2.4 Implementation

Systolic processors are attractive candidates for custom-VLSI implementation, due mainly to their low I/O requirement, regular structure and localized interconnection. Despite this, much recent effort has been spent trying to adapt the systolic approach to use off-the-shelf integrated circuits as well [2.7,8]. This effort has two primary objectives: first, to provide a vehicle for evaluating the many and varied systolic architectural and algorithmic alternatives; and second, to demonstrate that the benefits of a systolic processor are achievable even without going to custom integrated circuits. This section will address issues related to custom VLSI implementation as well as implementation by off-the-shelf, discrete components.

2.4.1 *Basic Systolic Cell Design*

A generic systolic cell for convolution-like computations needs a multiply-add unit, and the staging registers on input and output to permit each cell to function independently of its neighbors. The addition of a systolic control path and a cell memory makes this an excellent building block for a linear systolic array. To implement the two-stream, 1-D systolic array for the 2-D convolution, this basic cell must be modified to include the second

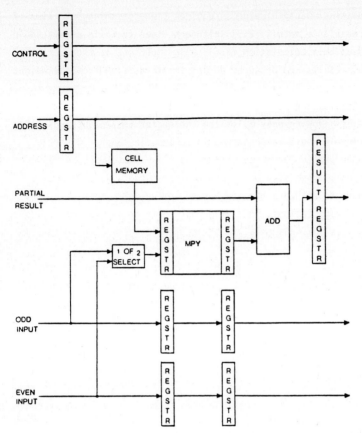

Fig.2.10. Basic two-stream systolic cell block diagram

systolic input path and the two-to-one multiplexer for stream selection. The resulting cell is diagrammed in Fig.2.10. Note that each input stream includes two registers to guarantee that it moves at half the rate of the output stream, as required in Sect.2.2.2.

The practicality of this design can be illustrated by examining a real problem, such as 4×4 cubic resampling. This is the inner loop for generalized spatial image warping, and is characteristic of the selected kernel applications described in Sect.2.3. The control variable, in this case, is the fractional position in the input image of each output pixel. Specifying this fractional position with 5-bit resolution in both the line and pixel directions will provide better than 0.1 pixel accuracy in the resampled image. A two-stream cell to accomplish this resampling could be constructed with 8-bit input paths and 12-bit weights. The 32×32 interpixel grid requires 1024 different weights in each cell memory, and a 10-bit systolic address path derived from the fractional pixel location. A 12-bit multiplier with a 24-bit partial-product adder will allow full precision to be maintained throughout the calculation.

As indicated in Sect.2.2, for higher-dimensional (more than 2-D) convolutions the only modification required to the basic two-stream cell is the addition of the requisite number of systolic input streams, along with the correspondingly sized multiplexer to select the proper data at each cell.

2.4.2 Implementation by Discrete Components

For a low-cost implementation of the basic systolic cell described above, the multiplier will be the limiting function on the cell speed. Using a TRW MPY-12HJ gives a worse-case cell computation rate of 110 nanoseconds. A standard Schottky-TTL, carry look-ahead adder and a high-speed 1K×4 static memory (e.g., Intel 2148) are compatible with this overall computation rate. This results in a cell having an arithmetic rate of greater than 18 Million Operations Per Second (MOPS), using only 24 integrated circuits and costing less than $250.

The complete 4×4 (16-cell), two-stream systolic interpolator would have an aggregate performance of 288 MOPS, or an output rate of 9 megapixels per second. Using a rather conservative printed circuit (PC) board density of 0.75 square inches per 16-pin IC, this entire processor would fit on two large PC boards. The total board area would be less than 500 square inches and the total material cost under $5000.

The systolic array cell just described can be used for all three convolutional classes outlined in Sect.2.3, except with true adaptive kernels. This class can be easily included by adding an extra data path to write to cell memory, thereby allowing the weights to be modified without using one of the operational data paths.

One important observation must be made with respect to adaptive kernels. At the transition from one weight set to another, the implication is that all the weights must change before the next computation cycle to guarantee that the weights used will be entirely from one set. For the systolic processor though, this is not the case. The contribution of each weight to the final convolution occurs in separate cells, one clock time apart. So, the weights may also be updated in this sequential manner. For a convolution using a $k \times k$ kernel, the required input bandwidth to the cell memory is no greater than the data input bandwidth, even though all k^2 weights must be replaced between computations. As a consequence of this, the adaptation rate is limited to once every k^2 cycles.

2.4.3 Custom VLSI Implementation by the PSC

An efficient way to implement the various flexibilities required by the weight selection and adaptation methods discussed above is to use a programmable processor chip that has both arithmetic and control facilities. The CMU Programmable Systolic Chip (PSC) [2.9,10] was designed for this purpose. The PSC is a high performance, special-purpose microprocessor intended to be used in large groups for the efficient implementation of a broad variety of systolic arrays. As depicted in Fig.2.11, the PSC building-block approach is to assemble many different types and sizes of systolic arrays from the same chip—by programming, each PSC can implement one systolic cell and many PSCs can be connected at the board level to build a systolic array.

This view was the starting point, in October of 1981, of the PSC project. The goal of the project has been the design and implementation of a prototype PSC in 4-micron nMOS which would demonstrate the benefits of the PSC approach, and in addition would itself constitute a cost-effective means for implementation of many systolic arrays. To ensure sufficient flexibility to cover a broad range of applications and algorithms, we chose an initial set of target applications for the PSC to support, including signal and image processing, error correcting codes, and disk sorting. The demands of these applications have resulted in the following design features:

- 3 eight-bit data input ports and 3 eight-bit data output ports;
- 3 one-bit control input ports and 3 one-bit control output ports;
- Eight-bit ALU with support for multiple precision and modulo 257 arithmetic;
- Multiplier-accumulator with eight-bit operands;
- 64-word by 60-bit writable control store;
- 64-word by 9-bit register file;
- Three 9-bit on-chip buses;
- Stack-based microsequencer.

The PSC has been fabricated and working functionally since March 1983. A system demonstration of the PSC is its implementation of a 1-D systolic array for 2-D convolution, based on the scheme of this paper. As of March 1984, a demonstration systolic array with nine PSCs is operational. The systolic array is connected to a SUN workstation. Due to its programmability and on-chip parallelism, we can program the PSC to implement all the operations of the cell of Fig.2.6b in one PSC instruction cycle, which should take no more than 200 ns assuming a state-of-the-art design for the PSC. This implies, for ex-

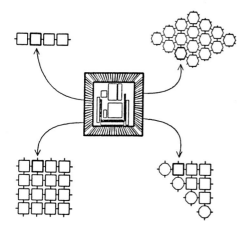

Fig.2.11. PSC: a building-block chip for a variety of systolic arrays

ample, that the 2-D convolution with 5×5 kernels can be implemented by 25 linearly connected PSCs, capable of producing one output pixel every 200 ns.

For large kernels, it is necessary to save partial results y_i in double or higher precision to ensure numerical accuracy of the computed results. As for the case of systolic arrays for 1-D convolution [2.1], this can be achieved most cost effectively by using a dual of the design in Fig.2.6, where partial results y_i stay in cells but the x_i and w_{ij} move from cell to cell in the same direction but at two speeds. With this dual design, high-precision accumulation can be effectively achieved by the on-chip multiplier-accumulator circuit, and the number of bits to be transferred between cells is minimized. We of course still need to transfer the final computed values of the y_i out of the array, but they, being the final results, can be truncated and transferred in single precision. It is generally preferable to have the w_{ij} rather than the x_i going through an additional register in each cell, since there are two x data streams but only one weight stream. Sometimes the communication cost for the weight stream can be totally eliminated if the register file of the PSC is large enough to hold a complete weight table.

2.5 Concluding Remarks

We have shown a regular and realizable systolic array architecture capable of performing multidimensional (n-D) convolutions with any kernel size k^n. This array can fully utilize k^n linearly connected cells, but more importantly, is sustained with a data input bandwidth which is independent of k and requires less than 2^{n-1} words per cell cycle. As a result, any convolution size (variation in k) can be handled with a fixed system input/output bandwidth capability, provided that k^n cells are used. To accommodate a larger value of k only requires that the existing systolic array be extended linearly with the appropriate number of identical cells.

An important characteristic of the proposed systolic array is that the data all flow in one direction. This property implies that the systolic array can be made fault tolerant with little overheads, and can be transformed automatically into one where cells are implemented with pipelined functional units, using results of a recent paper [2.11]. For instance, the cells of floating-point systolic arrays are typically implemented with multipliers and adders with three or more pipeline stages. It is therefore important to make effective use of this *second level of pipelining* provided by the cells themselves to increase the throughput of the systolic array further. A different two-level pipelined systolic array for multidimensional convolution was proposed previously [2.4], but that method requires a large buffer memory in every cell.

Multidimensional convolution and resampling represent just one class of computations suitable to systolic architectures. Recent research has shown that systolic architectures are possible for a wide class of compute-bound computations where multiple operations are performed on each data item in a repetitive and regular manner. This is especially true for much of the time-consuming "front-end processing" that deals with large amounts of data obtained directly from sensors in various signal- and image-processing application areas. To indicate the breadth of the systolic approach, the following is a partial list of problems for which systolic solutions exist.

a) Signal and Image Processing:
- FIR, IIR filtering
- 1-D, 2-D convolution and correlation
- 1-D, 2-D interpolation and resampling
- 1-D, 2-D median filtering
- discrete Fourier transforms
- geometric warping
- encoding/decoding for error correction

b) Matrix Arithmetic:
- matrix-vector multiplication
- matrix-matrix multiplication
- matrix triangularization (solution of linear systems, matrix inversion)

- QR decomposition (least-squares computation, covariance matrix inversion)
- singular value decomposition
- eigenvalue computation
- solution of triangular linear system
- solution of toeplitz linear systems

c) <u>Nonnumeric Applications:</u>
1) data structures
- stack, queue, priority queue
- searching, dictionary
- sorting
2) graph and geometric algorithms
- transitive closure
- minimum spanning trees
- connected components
- convex hulls
3) language recognizer
- string matching
- regular language recognition
4) dynamic programming
5) relational database operations
6) polynomial algorithms
- polynomial multiplication and division
- polynomial greatest common divisor
7) multiprecision integer arithmetic
- integer multiplication and division
- integer greatest common divisor
8) Monte Carlo simulation.

Several theoretical frameworks for the design of systolic arrays are being developed [2.11-14]. Extensive references for the systolic literature can be found in separate papers [2.1,15]. Major efforts have now been started to build systolic processors and to use them for large, real-life applications. Practical issues on the implementation and use of systolic array processors in systems are beginning to receive substantial attention. Readers can follow some of the references of this paper for further information on systolic arrays and their implementation.

References

2.1 H.T. Kung: Why Systolic Architectures? Computer Mag. **15**, 37-46 (January 1982)

2.2 H.T. Kung, C.E. Leiserson: Systolic Arrays (for VLSI), in *Sparse Matrix Proceedings 1978*, ed. by I.S. Duff and G.W. Stewart (Soc. for Industrial and Applied Mathematics, 1979) pp.256-282
A slightly different version appears in *Introduction to VLSI Systems*, ed. by C.A. Mead and L.A. Conway (Addison-Wesley, Reading, MA 1980) Sect.8.3

2.3 H.T. Kung, S.W. Song: A Systolic 2-D Convolution Chip, in *Multicomputers and Image Processing: Algorithms and Programs*, ed. by K. Preston, Jr. and L. Uhr (Academic, New York 1982) pp.373-384

2.4 H.T. Kung, L.M. Ruane, D.W.L. Yen: Two-Level Pipelined Systolic Array for Multidimensional Convolution. Image and Vision Computing **1**, 30-36 (1983)
An improved version appears as a CMU Computer Science Department technical report November 1982

2.5 R.A. Monzingo, T.W. Miller: *Introduction to Adaptive Arrays* (Wiley, New York 1980)

2.6 W.M. Gentleman, H.T. Kung: Matrix Triangularization by Systolic Arrays, in *Real-Time Signal Processing IV*, SPIE Symp. **298**, 16-26 (Soc. Photo-Optical Instrumentation Engineers, 1981)

2.7 J.J. Symanski: *NOSC Systolic Processor Testbed*. Technical Report NOSC TD 588, Naval Ocean Systems Center (June 1983)

2.8 D.W.L. Yen, A.V. Kulkarni: Systolic Processing and an Implementation for Signal and Image Processing. IEEE Trans. C-**31**, 1000-1009 (1982)

2.9 A.L. Fisher, H.T. Kung, L.M. Monier, Y. Dohi: Architecture of the PSC: A Programmable Systolic Chip, in *Proc. 10th Annual Intern. Symp. Computer Architecture* (June 1983) pp.48-53

2.10 A.L. Fisher, H.T. Kung, L.M. Monier, H. Walker, Y. Dohi: Design of the PSC: A Programmable Systolic Chip, in *Proc. 3rd Caltech Conf. on Very Large Scale Integration*, ed. by R. Bryant (Computer Science Press, Rockville, MD 1983) pp.287-302

2.11 H.T. Kung, M. Lam: Fault-Tolerance and Two-Level Pipelining in VLSI Systolic Arrays, in *Proc. Conf. Advanced Research in VLSI* (Artech House, Inc., Cambridge, MA, January 1984) pp.74-83

2.12 H.T. Kung, W.T. Lin: An Algebra for VLSI Computation, in *Elliptic Problem Solvers II*, ed. by G. Birkhoff and A.L. Schoenstadt (Academic, New York 1983). Proc. Conf. on Elliptic Problem Solvers, January 1983

2.13 C.E. Leiserson, J.B. Saxe: Optimizing Synchronous Systems. J. VLSI and Computer Syst. **1**, 41-68 (1983)

2.14 U. Weiser, A. Davis: A Wavefront Notation Tool for VLSI Array Design, in *VLSI Systems and Computations*, ed. by H.T. Kung, R.F. Sproull, and G.L. Steele, Jr. (Computer Science Press, Rockville, MD 1981) pp.226-234

2.15 A.L. Fisher, H.T. Kung: Special-Purpose VLSI Architectures: General Discussions and a Case Study, in *VLSI and Modern Signal Processing* (Prentice-Hall, Reading, MA 1982)

3. VLSI Arrays for Pattern Recognition and Image Processing: I/O Bandwidth Considerations

T.Y. Young and P.S. Liu

Very Large Scale Integration (VLSI) computing arrays for pattern recognition and image processing have received increasing attention in recent years. The computation speed of an array is often limited by the I/O bandwidth of the host system or the VLSI array. Reconfiguration techniques are used to restructure a computing array, so that successive functional computation steps can be carried out without the data leaving the array. Computation time is reduced, since with reconfiguration the limited I/O bandwidth affects only the first and last phases of the necessary computations.

Covariance matrix inversion and matrix multiplication are essential computation steps in statistical pattern recognition. A new orthogonally connected triangular array for L-U decomposition of a covariance matrix is presented. A square array is reconfigured into a triangular array and vice versa by modifying the communication structure along the central diagonal of the square array. It is shown that generation of the covariance matrix from sample vectors and linear and quadratic discriminant functions can be computed on the same array. The image-processing array is capable of performing DFT and various spatial domain operations. The effect of limited I/O bandwidth on the computation speed of a multiple-chip partitioned array is studied.

3.1 Background

The VLSI array architecture for concurrent computation [3.1-6] has generated substantial interest in recent years; and its applications to signal processing [3.7,8], image processing [3.9-14], and pattern recognition [3.14-18] have been discussed. Many of these computing structures use the systolic array approach, which incorporates multiprocessing and data pipelining externally and internally. Such an array may be attached or interfaced to a host system or systems as a special-purpose device.

The computation speed of a VLSI array may be limited by the bandwidth of the host system or the VLSI array, whichever is smaller. The bandwidth is directly related to the data rate and number of the host system's I/O lines or the number of the computing array's I/O pins. With limited I/O bandwidth or capacity, it may not be possible in some cases to pipe sufficient data into the computing array to sustain continuously the activities of all the processing elements that can be possibly fabricated into the VLSI computing array. With this situation, the design of the computing array should aim at minimizing the adverse effect of limited I/O capacity and preserving the large data storage and parallel computation advantages offered by VLSI computing arrays.

Two design approaches may be used to alleviate the adverse effect caused by limited I/O bandwidth. The first approach is to make sure that the available I/O pins of the array are utilized efficiently to minimize total computation time. In a recent paper [3.6], we examined three configurations for interfacing and controlling a matrix multiplication array. A properly chosen configuration can significantly reduce the computing time of the multiplication array.

A second design approach is to restructure an array using reconfiguration techniques. In pattern analysis and image processing, an operation often consists of several computation steps. If several chips are used, one for each computation step, the I/O limitation of VLSI chips would affect every computation step. With a reconfigurable array, the limited I/O bandwidth affects the execution time only during the first and last computation steps; at all other times of functional computations, the array is compute bound. Thus, the overall execution time can be reduced significantly by reconfiguration. A reconfigurable array also offers the flexibility of using the same array for several related operations.

A systolic array is usually designed for a single functional computation such as matrix multiplication. It is shown in this chapter that a systolic array can be modified for different functional computations while retaining the overall simplicity of the array. Reconfiguration of the array is accomplished primarily by diverting the data streams.

It is assumed that pattern classification is based on a linear or quadratic discriminant function. Thus during the learning phase, it is necessary to compute the covariance matrix from sample vectors and to calculate matrix inversion. During the classification phase, the computations involved are matrix multiplication, matrix-vector multiplication, and vector dot product. An orthogonally connected array can be reconfigurated to perform all these functional computations.

The special-purpose image-processing array is designed for high-throughput operations such as image enhancement and segmentation. The array is capable of computing the discrete Fourier transform (DFT) and various spatial domain operations. Since a digital image often consists of a very large number of pixels, it may be necessary to partition the array into subarrays, with each array fabricated on a VLSI chip. The effect of interchip I/O limitation is examined.

3.2 Arrays for Matrix Operations

Matrix multiplication and covariance matrix inversion are commonly used in pattern analysis and image processing [3.19]. It is shown in this section that an orthogonally connected, reconfigurable square array is capable of performing matrix multiplication and inversion. The data paths leading into the processing elements (PE's) along the diagonal of the square array can be reconfigurated. With reconfiguration, the various steps required for covariance matrix inversion can be carried out without the data leaving the array.

3.2.1 Matrix Multiplication

Consider the multiplication of two $n \times n$ matrices,

$$Z = AB \ . \tag{3.1}$$

Assuming the matrix **A** is already inside the array, **B** can be piped in to interact with **A** to produce the multiplication result as shown in Fig.3.1. During each computation cycle, all the cells in the array perform the same multiplication step: each processing element PE_{ij} takes the sum of partial products from its left neighbor and adds it to its partial product of $a \times b$, and then passes the sum to its right neighbor for the next multiplication step [3.12,13]. For each multiplication step, the data streams of **B** are piped one row deeper into the array. It is noted that similar to other systolic arrays, the input and output streams are skewed.

Without I/O limitation, the computation time is $(4n-2)$ units, including n units of matrix **A** loading time and $(3n-2)$ units of multiplication time. A PE, once activated, performs the necessary multiplication at every cycle time, until no more data enters the PE. At the peak of computation, about 75% of the n^2 PE's compute simultaneously.

Three configurations for the interfacing and controlling of the multiplication array under the constraint of limited I/O bandwidth have been studied

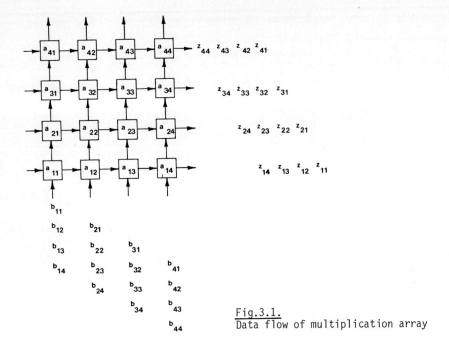

Fig.3.1.
Data flow of multiplication array

[3.6]. The three configurations are multiplexing loading, processor row loading, and processor column group loading. It was shown that a properly chosen configuration could reduce the computing time of the array by 25%. Similar analysis can be made for the study of I/O limitation and control and interfacing schemes of other simple computing arrays.

3.2.2 L-U Decomposition

A reconfigurable, orthogonally connected square array is used to compute the inverse of a symmetric nonsingular covariance matrix $R = [r_{ij}]$. The first step of the computation is to decompose it into an upper triangular matrix $U = [u_{ij}]$ and a lower triangular matrix $L = [\ell_{ij}]$. To decompose the covariance matrix, we use the following recurrence procedure:

$$c_{ij}^{(1)} = r_{ij} ,$$

$$c_{ij}^{(k+1)} = c_{ij}^{(k)} - \ell_{ik} u_{kj} ,$$

$$\ell_{ik} = \begin{cases} 0 & \text{if } i < k , \\ 1 & \text{if } i = k , \\ c_{ik}^{(k)} u_{kk}^{-1} & \text{if } i > k , \end{cases}$$

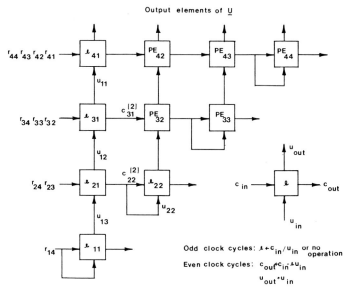

Fig.3.2. Triangular array for L-U decomposition of a covariance matrix

$$u_{kj} = \begin{cases} 0 & \text{if } k > j, \\ c_{kj}^{(k)} & \text{if } k \leq j. \end{cases} \quad (3.2)$$

The operations required of each processing element and data movement of the array are shown in Fig.3.2, assuming a 4×4 covariance matrix. The PE has 2 inputs c_{in} and u_{in} and 2 outputs c_{out} and u_{out} which are buffered or latched internally, and it has an internal storage register ℓ. Each PE operates in two modes, the multiplication mode and the division mode. It is assumed that each operation takes one unit of time to complete. The multiplication mode allows the PE's in the array to implement the recurrence $c_{ij}^{(k+1)} = c_{ij}^{(k)} - \ell_{ik} u_{kj}$ in the decomposition and at the same time move u_{kj} upward. The division mode allows a PE to calculate $\ell_{ik} = c_{ik}^{(k)} u_{kk}^{-1}$ when necessary.

The skewed columns of the matrix **R** are piped into the triangular array in parallel as inputs. The input data streams are piped one column deeper into the array every two units of time. The array goes through alternate division and multiplication operations. For each multiplication operation, all PE's are activated. For each division operation, however, only some PE's with the proper input data at that time will be allowed to do division and store the result in the internal register ℓ. In general, and starting at the

29

Table 3.1. Computation sequence of ℓ_{ik}

Time interval	ℓ_{ik} computed	PE involved
T_1	ℓ_{11}	PE_{11}
T_3	ℓ_{21}	PE_{21}
T_5	ℓ_{31}	PE_{31}
T_7	ℓ_{22}, ℓ_{41}	PE_{22}, PE_{41}
T_9	ℓ_{32}	PE_{32}
T_{11}	ℓ_{42}	PE_{42}
T_{13}	ℓ_{33}	PE_{33}
T_{15}	ℓ_{43}	PE_{43}
T_{19}	ℓ_{44}	PE_{44}

bottom PE_{11}, the ℓ_{ik} elements are computed during odd time intervals. Table 3.1 shows the time and place at which each ℓ_{ik} is computed.

The computation pattern for ℓ_{ik} can be generalized. Each computation of ℓ_{ik} is delayed successively by 2 units of time for the PE's within each column of the array and by 4 units of time for the PE's within each row of the array. A computed ℓ_{ik} is stored inside its PE and used to compute subsequent $c_{ij}^{(k)}$'s and u_{kj}'s. Control signals could be piped and distributed with proper delay through the processing elements themselves, or the control timing could be customized and built into each PE individually. It can be deduced that decomposition of any $n \times n$ covariance matrix takes $(6n-3)$ units of time. With $n(n+1)/2$ PE's used in the array, the space-time product is approximately $3n^3$. Equation (3.2) is based on Crout's method [3.20] which does not take into consideration the symmetry of the covariance matrix. The basic triangular array structure can be modified [3.6] to implment Cholesky's decomposition method for symmetric matrices. The number of time units required is reduced by half; however, the unit cycle time is longer in this case, since it is limited by the computation of square roots. We choose Crout's method because the unit cycle time is limited by multiplication or division, and because the same type of PE's can be used for other functional computations.

3.2.3 A Reconfigurable Array for Matrix Inversion

As shown in Fig.3.3, the orthogonally connected square array is reconfigurated along the diagonal to become an upper triangular array for L-U decomposition. The next step for covariance matrix inversion is the computation of $V = U^{-1}$ using a lower triangular array. The processing operations of the

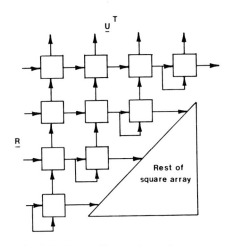 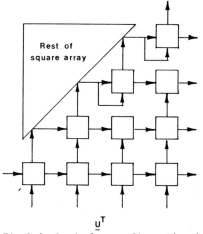

Fig.3.3. Reconfiguration of a square array into an upper triangular array for L-U decomposition

Fig.3.4. Gradual reconfiguration into a lower triangular array for the computation of $V = U^{-1}$

PE's in this step is very similar to that in L-U decomposition with different data as inputs. During odd clock cycles, we have $v \leftarrow c_{in}/u_{in}$ or no operation. During even clock cycles, we have $c_{out} \leftarrow c_{in} - vu_{in}$ and $u_{out} \leftarrow u_{in}$. As soon as u_{11} and subsequent u_{ij}'s leave the top of the array, they are piped back without delay to the bottom of the array for the computation of V. The array is gradually reconfigurated into the lower triangular array as shown in Fig.3.4, with the PE_{ii}'s on the diagonal included in the triangular array. The matrix V remains in the triangular array after the computation.

Step 3 is the computation of the inverse of L, which is denoted by $M = [m_{ij}]$, and since R is symmetric $m_{ji} = v_{ij}u_{jj}$. In this step, duplicated values of v_{ij} in each processor column are rotated together four times along each column of the square array. As a v_{ij} passes through PE_{ii}, $m_{ji} = v_{ij}u_{jj}$ is calculated and shifted upward as replacement for v_{ij}. At the end, both v_{ij} and m_{ji} reside in the same PE_{ij}.

In the final step from within the array, the matrix M is piped to the top and then into the array again through the bottom. Piping of each processor column is delayed by one unit time. Moreover, M interacts with V still inside the array, and multiplication steps are carried out as defined in the multiplication array. The output is therefore $U^{-1}L^{-1}$ which is the inverse of R. At the end of computation, R^{-1} may be piped out or returned to the array.

3.2.4 Performance Analysis

Let us first consider the case that the four steps of matrix inversion are performed for an $n \times n$ matrix R using four separate special-purpose chips. With sufficiently large I/O bandwidth, the execution time of step 1 is $(6n-3)$ units, which is also the time required for step 2. It can be seen from above discussions that overlap I/O and computations exist between the two arrays; thus with overlaps accounted for, the execution time of step 1 and step 2 is $(8n-2)$. The time required for step 3 and step 4 is n and $(4n-2)$ units respectively. It is noted that step 3 cannot start until all results from step 2 become available, and step 4 cannot start until step 3 is completed. The total execution time is, therefore, $(13n-4)$ units.

With limited I/O bandwidth, we assume that only f pins are available for either input or output purposes for every array, and $f < n$. Then, the total execution time is $n(13n-4)/f$ units.

Using the reconfigurable array, the limited I/O bandwidth affects the first and last steps only. Thus, the execution time of step 1 and step 2 is, under the limitation of f input pins, $n(4n-2)/f+4n$. Step 4 requires $n(4n-2)/f$ time units, since it is limited by the f output pins. The total execution time for the reconfigurable array is $n(8n-4)/f+5n$ units. Under severe pin limitations, n/f becomes a large number. The reconfigurable array takes approximately $8n^2/f$ units of time to compute, compared to $13n^2/f$ units for separate arrays. The required computation time is reduced by 38%.

3.3 Arrays for Pattern Analysis

In statistical pattern classification [3.19], a set of n measurements is regarded as an n-dimensional vector in a vector space. The vector space is divided into several regions based on statistical analysis or decision theory, with each region corresponding to one class of patterns. *Hwang* and *Su* [3.15] discussed VLSI arrays for the computation of Fisher's linear discriminant, using a partitioned matrix approach. *Liu* and *Fu* [3.16] considered VLSI architecture for minimum distance classifiers, an important subclass of linear classifiers.

Pattern classification can be divided into two phases, the learning phase and the classification phase. If the probability distributions of the classes are known except for some parameter values to be estimated, Bayes decision rules may be adopted, and the resulting discriminant functions are essentially log-likelihood functions. Often the probability distributions are assumed Gaussian; then the discriminant functions are linear or quadratic depending on whether or not the covariance matrices of the different classes are identical. During the learning phase, covariance matrices and mean vectors are estimated from sample vectors of known classifications. Discriminant functions are obtained, and new samples of unknown classifications are classified using the discriminant functions.

3.3.1 Discriminant Functions

Let **x** be an n-dimensional column vector. With a Gaussian distribution, the discriminant function for class k may be expressed as [3.19]

$$D_k(\mathbf{x}) = -2 \log p_k(\mathbf{x})$$

$$= (\mathbf{x} - \boldsymbol{\mu}_k)^T R_k^{-1} (\mathbf{x} - \boldsymbol{\mu}_k) + \log(2\pi)^n |R_k| \quad , \qquad (3.3)$$

where R_k and $\boldsymbol{\mu}_k$ are covariance matrix and mean vector, respectively. This is a quadratic discriminant function, and the pattern vector **x** will be assigned to class k, if $D_k(\mathbf{x})$ is minimum among the values of discriminant functions of all classes.

If the covariance matrices are identical, the pairwise discriminant function for class j and class k is linear,

$$D_{jk}(\mathbf{x}) = D_j(\mathbf{x}) - D_k(\mathbf{x})$$

$$= 2\mathbf{x}^T R^{-1} (\boldsymbol{\mu}_k - \boldsymbol{\mu}_j) - \boldsymbol{\mu}_k^T R^{-1} \boldsymbol{\mu}_k + \boldsymbol{\mu}_j^T R^{-1} \boldsymbol{\mu}_j \quad . \qquad (3.4)$$

Class j is preferred if $D_{jk}(\mathbf{x}) < 0$.

To estimate the covariance matrices and the mean vectors, we assume that $\mathbf{x}_1, \mathbf{x}_2, \ldots, \mathbf{x}_m$ belong to the same class k, and omit the subscript k for simplicity of notation. Then

$$\boldsymbol{\mu} = \frac{1}{m} \sum_{i=1}^{m} \mathbf{x}_i$$

$$R = \frac{1}{m} \sum_{i=1}^{m} \mathbf{x}_i \mathbf{x}_i^T - \boldsymbol{\mu} \boldsymbol{\mu}^T \quad . \qquad (3.5)$$

3.3.2 Reconfigurable Array Processor

It is clear from the above discussions that statistical pattern classification involves matrix and vector operations. A pattern analysis array processor is shown in Fig.3.5a. It consists of a square array and a linear array, with the

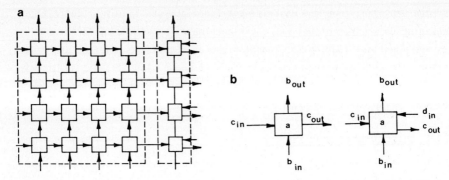

Fig.3.5. (a) Pattern analysis array processor, and (b) processing elements

n horizontal outputs of the square array connected to the inputs of the linear array. The array processor is reconfigurable in the sense that data paths can be changed by modifying the communication structure between the two arrays and/or the structures among the PE's. A local buffer memory may be needed for temporary storage and I/O purposes.

Each PE of the square array has two inputs c_{in} and b_{in} and two outputs c_{out} and b_{out}, as shown in Fig.3.5b, and a PE may have several internal storage registers. It can perform three types of operations: (i) data transfer and loading, (ii) multiplication- and-addition (or subtraction), and (iii) division. There may be several slightly different operations within an operation type. For example, matrix multiplication requires $c_{out} \leftarrow c_{in} + ab_{in}$, while L-U decomposition needs multiplication-and-subtraction as indicated by (3.2). In addition, each PE can compute $a \leftarrow a + c_{in} b_{in}$, which is needed for the computation of covariance matrices. Simple addition, subtraction and multiplication are regarded as special cases of (ii). For the linear array, each PE has three inputs and two outputs. Its operations are very similar to that of the PE's of the square array.

Data streams are piped into the array in a skewed manner. A multiplication-and-addition or division operation is executed in one unit of cycle time. Data transfer operations can be executed simultaneously with the other operations.

With skewed data streams as inputs, individual PE or groups of PE's of the square array are activated in the following sequence: PE_{11}, (PE_{12}, PE_{21}), $(PE_{31}, PE_{22}, PE_{31}), \ldots, PE_{nn}$. Thus, operation commands for covariance matrix calculations and pattern classification computations can be piped or distributed to the PE's starting from PE_{11}. One way to distribute the operation commands is shown in Fig.3.6 for a 4×4 array. Each PE labeled k obtains its operation commands through another PE labeled (k-1). The labeling scheme is

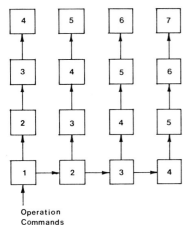

Fig.3.6. Distribution paths of operation commands for the square array

derived from the sequence by which the PE's are activated. A similar scheme may be used for the linear array.

3.3.3 Computation of Covariance Matrices

The covariance matrix and the mean vector of a certain class is calculated from a set of m sample vectors belonging to that class. Let

$$X = [x_1, x_2, \ldots, x_m] \quad . \tag{3.6}$$

The block diagram of the computation using the pattern analysis array processor is shown in Fig.3.7.

The skewed input data X are piped into the array from two directions. The computation can be divided into three steps. In the first step, PE_{jk} of the square array computes $a \leftarrow a + x_{ji}x_{ki}$ as the data streams arrive, and PE_j of the linear array calculates $a \leftarrow a + x_{ji}$. After the data streams of X passed through a PE, the content of the register a is divided by m, so that $(\Sigma\ x_{ji}x_{ki})/m$ is stored in PE_{jk} of the square array. The vector m1, after

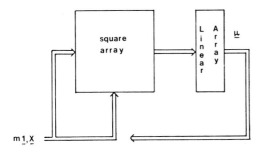

Fig.3.7. Computation of covariance matrices and mean vectors

passing through the square array, enters the linear array for the computation of $\mu_j = (\Sigma_i x_{ji})/m$. The last step is to pipe μ back into the square array from two directions and to compute at PE_{jk} $a \leftarrow a - \mu_j \mu_k$. Thus, after the last step of computation, we have the matrix R stored in the square array, and the vector μ in the linear array. Since the input data streams are skewed, the total computation time is $(2n+m+1)$ units.

3.3.4 Computation of Discriminant Functions

The computation of R^{-1} using a reconfigurable array has been discussed in Sect.3.2. It is noted that

$$|R| = |L||U| = 1 \times |U| = \prod_{j=1}^{n} u_{jj} \quad . \tag{3.7}$$

During the computation of U^{-1}, the matrix U leaves the square array from the top of the array, and it is piped back without delay to the bottom of the array. Thus, it is clear that the diagonal elements of U can be diverted and fed into the linear array for simultaneous computation of the determinant $|R|$.

The block diagram for computing the quadratic term of the quadratic discriminant function is shown in Fig.3.8, where X represents m sample vectors being classified. The quadratic term may be expressed as $(x-\mu)^T(R^{-1}x - R^{-1}\mu)$. With R^{-1} stored in the square array, the array essentially performs matrix multiplication. Its skewed outputs, $R^{-1}\mu$ and $R^{-1}x$, are piped into the linear array, and at the same time, the skewed data streams, μ and X, having passed through the square array, enter the linear array from the other side. The vectors μ and $R^{-1}\mu$ are stored in internal registers of the linear array, before the array executes the appropriate multiplication-and-subtraction operations. The computation time is $(2n+m)$ units.

Since there are several classes of patterns, local memory is needed to store the data X and the several covariance matrices and mean vectors. Comparison of the several $D_k(x)$ may be carried out in a special linear 'compare' processor [3.16] or in the host computer.

The computation of pairwise linear discriminant functin is shown in Fig. 3.9. The data flow pattern is very similar to the flow pattern in the computation of the quadratic term. The linear array first computes $\mu_j^T R^{-1} \mu_j$ and $\mu_k^T R^{-1} \mu_k$, and then with μ_j and μ_k already stored in the array, it computes $x^T R^{-1}(\mu_k - \mu_j)$. The computation time is $(2n+m+1)$ units. It is noted from (3.4) that it may not be necessary to compute $R^{-1}x$, and an alternate scheme can be developed by modifying the data flow pattern; however, the computation

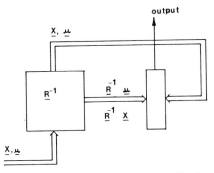
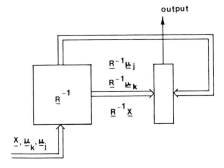

Fig.3.8. Computation of $(\mathbf{x}-\boldsymbol{\mu})^T\mathbf{R}^{-1}(\mathbf{x}-\boldsymbol{\mu})$

Fig.3.9. Computation of pairwise linear discriminant function

time remains the same with the PE's of the square array idle most of the time.

3.4 Image-Processing Array

Many operations in image processing are performed repeatedly over a large number of pixels of an image, and computation time can be reduced significantly by VLSI parallel processing. We consider here VLSI arrays for high throughput image operations such as image enhancement and segmentation.

The gray levels of the $n \times n$ pixels in an image can be represented by a matrix $\mathbf{X} = [x_{ij}] = [x(i,j)]$. An $n \times n$ orthogonally connected square array will be used for spatial domain and frequency domain operations. When the image size is large, it may be necessary to use a number of chips to accommodate the large number of pixels, and orthogonal connection results in one of the simplest structures for intra- and interchip connections.

3.4.1 Computation of DFT

The two-dimensional discrete Fourier transform may be expressed in matrix form,

$$\mathbf{Y} = \mathbf{A}\,\mathbf{X}\,\mathbf{B} \ . \tag{3.8}$$

Assuming $n = 4$, the matrices $\mathbf{A}$ and $\mathbf{B}$ become

$$\mathbf{A} = \mathbf{B} = \begin{bmatrix} 1 & 1 & 1 & 1 \\ 1 & \omega & \omega^2 & \omega^3 \\ 1 & \omega^2 & \omega^4 & \omega^6 \\ 1 & \omega^3 & \omega^6 & \omega^9 \end{bmatrix} \tag{3.9}$$

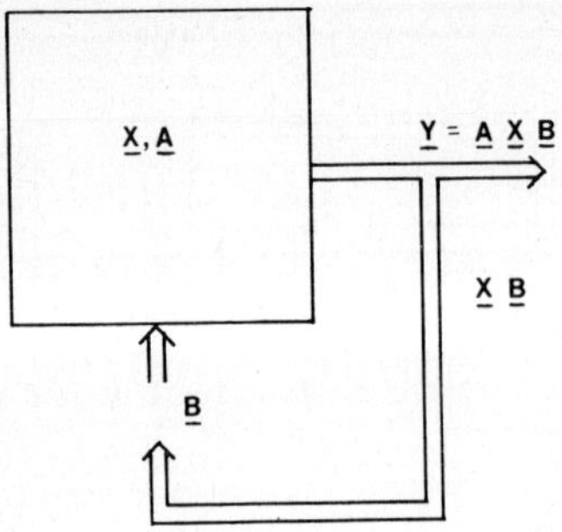

Fig.3.10. Computation of DFT using square matrix multiplication array

where $\omega = \exp(-j2\pi/n)$.

The architecture for DFT using matrix multiplication array has been studied by us [3.12,13], and a block diagram is shown in Fig.3.10. It is assumed that both **X** and **A** have been loaded into the square array. With the matrix **B** piped in from below, it interacts with X to form a matrix product **X B**, which in turn is fed back and interacts with the stored **A** to form the DFT, **Y** = **A X B**. Since **B** = **A**, we could use as input **B** the matrix **A** already stored in the square array.

The inverse DFT can be calculated in a similar manner. Thus, with a frequency domain filtering function loaded into the array, two-dimensional image filtering can be implemented.

3.4.2 Spatial Domain Operations

Template matching is a high throughput image operation. By selecting the weighting coefficients properly, the templates may be used for various spatial domain tasks, including line detection, edge detection, image smoothing, and image sharpening. To match a 3×3 template $W = [w(k,\ell)]$ we nedd to compute

$$\hat{x}(i,j) = \sum_{k=-1}^{1} \sum_{\ell=-1}^{1} w(k,\ell) x(i+k, j+\ell) \qquad (3.10)$$

and then compare the result with a predetermined decision threshold.

There are several possible approaches to VLSI implementation of template matching: *Kung* [3.9] suggested the use of (n-2) kernel cells, each kernel cell consisting of nine basic processing cells to form three linear arrays, plus a row interface cell to add the results emanated from the linear arrays. Three data streams enter a kernel cell, two of which exit from the cell after necessary computations, and then enter the next kernel cell. The communication structure in this case is incompatible with the n × n square array structure needed for frequency domain operations.

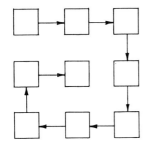

Fig.3.11. Data path for a 3×3 template matching computation method

The authors [3.12,13] proposed a matrix multiplication approach to template matching. As many as four matrix multiplication operations are needed for a 3×3 template. The values of the matrix elements must be precomputed from the given weighting coefficients, possibly at the host computer.

We now discuss a method which is based on a combination of systolic and cellular array concepts. The n×n square array is conceived as overlapping templates. The basic computation scheme is shown in Fig.3.11. With 3×3 templates, the (i,j)-th template computation involves 9 PE's in the following order: $PE_{i-1,j-1}$ - $PE_{i-1,j}$ - $PE_{i-1,j+1}$ - $PE_{i,j+1}$ - $PE_{i+1,j+1}$ - $PE_{i+1,j}$ - $PE_{i+1,j-1}$ - $PE_{i,j-1}$ - $PE_{i,j}$. All n^2 PE's perform multiply-and-add operations at the same time and it takes only 9 units of time to complete the template matching computation of the image. Each coefficient is presented to all PE's simultaneously, and all PE's are required to be active right from the beginning. This implies broadcasting of weighting coefficients and possibly control commands to the PE's. If the control algorithm is built into each PE with some kind of handshake synchronization among them, then broadcasting of control may not be necessary. An alternative approach is to pipe the coefficients through each row; the assumption is that the unit data transfer time from one PE to another is much shorter than the cycle time of the arithmetic operations so that the total execution time will not be dominated by the coefficient loading time.

With this last method of VLSI template matching, data paths between a pair of PE's must be bidirectional. The method is conceptually simple, and it can be easily generalized to other sizes of templates. As noted before, the same square array can be used for frequency domain operations by changing data flow patterns.

3.4.3 Partitioned Array—I/O Considerations

When n×n PE's can be fabricated into a single chip, I/O bandwidth or number of I/O pins affects the initial data loading time and the part of computation time that overlaps with data output from the chip. As images can have up to 512×512 pixels or more, it is likely that more than one chip is required and that the array needs to be partitioned into subarrays. Each subarray consists of $n_1 \times n_2$ PE's, and $n_1 n_2 < n^2$ so that the subarray can be fitted into a single chip. Assuming a chip with fixed silicon area for integrated circuits, and depending on the integration density, limited interchip (interpackage) I/O bandwidth or I/O pins may have an adverse effect on the computation speed of the array.

Let I^2 be the maximum number of PE's that can be fabricated on a chip, $g = g(I)$ be the number of I/O pins required to sustain the computation of I^2 PE's continuously in a chip, and f be the maximum number of I/O pins that can be put on a chip package. It is obvious that as long as $f \geq g$, the square array consisting of multiple chips does not suffer from limited I/O bandwidth.

When $f < g$, there are two extreme choices. The first one is to maintain array speed with $n_1 \times n_2$ PE's per chip such that $(n_1 + n_2) \leq 2If/g$. With this approach, part of the silicon area of the chip is not utilized, and the number of chips required may be large. The other extreme choice is to use a minimum number of chips at reduced speed by multiplexing the I/O pins. With $n_1 = n_2 = I$, the effective computation time in this case is g/f of that in the first choice.

In general, the effective computation time may be expressed as

$$t/t_a = (n_1 + n_2)g/(2fI) \;, \quad 2If/g \leq (n_1 + n_2) \leq 2I \;, \qquad (3.11)$$

where t_a is the array computation time without I/O limitation. It is noted that for a given value of $(n_1 + n_2)$, $n_1 n_2$ is maximum if $n_1 = n_2$, and hence $(n_1 + n_2) \leq 2I$ in (3.11) implies $n_1 n_2 \leq I^2$. The number of chips required is $n^2/(n_1 n_2)$.

3.5 Conclusions

We have studied the effect of limited I/O bandwidth on the computation speed of VLSI arrays. For simple operations such as matrix multiplication, computation time can be reduced significantly by selecting an appropriate configuration for interfacing and controlling the array.

Reconfiguration techniques are used to restructure a pattern analysis array so that it can carry out the successive phases of required computations without the data leaving the array. Computation time is reduced, since with the reconfigurable array the limited I/O bandwidth affects only the first and last phases of the necessary computations. The array can be used to generate covariance matrices, compute matrix inversion, and calculate linear and quadratic discriminant functions for pattern recognition.

The image-processing array is designed for both spatial domain and frequency domain operations. Because of the large number of pixels in an image, it is likely that the array will be partitioned and implemented on multiple chips. The effect of limited interchip I/O bandwidth on computation speed is examined.

References

3.1 H.T. Kung, C.E. Leiserson: Algorithms for VLSI Processor Arrays, in *Introduction to VLSI Systems*, ed. by C.A. Mead, L. Conway (Addison-Wesley, Reading, MA 1980) pp.271-292
3.2 L. Johnson: VLSI Algorithms for Doolittle's Crout's and Cholesky's Methods, in *Proc. IEEE Intern. Conf. on Circuits and Computers* (1982) pp.372-376
3.3 S.Y. Kung, K.S. Arun, R.J. Gal-Ezer, D.V. Bhaskar Rao: Wavefront Array Processor: Language, Architecture and Applications. IEEE Trans. C-**31**, 1054-1066 (1982)
3.4 H.M. Ahmed, J.M. Delsome, M. Morf: Highly Concurrent Computing Structure for Matrix Arithematic and Signal Processing. Computer Magazine **15**, 65-82 (1982)
3.5 K. Hwang, Y.-H. Cheng: Partitioned Matrix Algorithm for VLSI Arithmetic Systems. IEEE Trans. C-**31**, 1215-1224 (1982)
3.6 P.S. Liu, T.Y. Young: VLSI Array Design Under Constraint of Limited I/O Bandwidth. IEEE Trans. C-**32** (1983)
3.7 E.E. Swartzlander, Jr.: VLSI Technology for Signal Processing, in *Proc. Government Microcircuit Applications Conference*, Vol.7 (1978) pp.76-79
3.8 J.M. Speiser, H.J. Whitehouse, K. Bromley: Signal Processing Applications for Systolic Arrays. Naval Ocean System Center, San Diego, CA 92152, USA
3.9 H.T. Kung: Special Purpose Devices for Signal and Image Processing: An Opportunity in VLSI, Tech. Rep. CS-80-132 (Dept. Computer Sci. Carnegie-Mellon Univ., Pittsburgh, PA 1980)
3.10 P. Narendra: VLSI Architecture for Real Time Image Processing. Digest of Papers, COMPCON 81 (1981) pp.303-306
3.11 S.L. Tanimoto: An Image Processor Based on an Array of Pipelines. Proc. Workshop on Comp. Arch. for Pattern Analysis and Image Database Management (1981) pp.201-208
3.12 T.Y. Young, P.S. Liu: VLSI Arrays and Control Structure for Image Processing. Proc. Workshop on Comp. Arch. for Pattern Analysis and Image Database Management (1981) pp.257-264
3.13 P.S. Liu, T.Y. Young: VLSI Array Architecture for Picture Processing, in *Picture Engineering*, ed. by K.S. Fu, T. Kunii, Springer Ser. Inform. Sci., Vol.6 (Springer, Berlin, Heidelberg, New York 1982) pp.171-186
3.14 T.Y. Young, P.S. Liu: Impact of VLSI on Pattern Recognition and Image Processing, in *VLSI Electronics: Microstructure Science*, Vol.4, ed. by N.G. Einspruch (Academic, New York 1982) pp.319-360
3.15 K. Hwang, S.P. Su: A Partitioned Matrix Approach to VLSI Pattern Classification, in *Proc. Workshop on Comp. Arch. for Pattern Analysis and Image Database Management* (1981) pp.168-177

3.16 H.H. Liu, K.S. Fu: VLSI Algorithm for Minimum Distance Classification. Proc. IEEE Intern. Conf. on Computer Design: *VLSI in Computers* (1983)
3.17 K.H. Chu, K.S. Fu: VLSI Architecture for High-Speed Recognition of Context-Free Language. Proc. 9th Annual Symp. on Computer Architecture (1982) pp.43-49
3.18 P.S. Liu, T.Y. Young: VLSI Arrays with Limited I/O Bandwidth for Pattern Analysis. Proc. Workshop on Comp. Arch. for Pattern Analysis and Image Database Management (1983) pp.2-9
3.19 T.Y. Young, T.W. Calvert: *Classification, Estimation and Pattern Recognition* (Elsevier, New York 1974)
3.20 J.R. Westlake: *A Handbook of Numerical Matrix Inversion and Solution of Linear Equations* (Wiley, New York 1968)

Part II

VLSI Systems for Pattern Recognition

4. VLSI Arrays for Minimum-Distance Classifications

Hsi-Ho Liu and King-Sun Fu

Minium-distance classification (MDC) is a popular technique in statistical pattern recognition, clustering and many other applications. A reference vector, i.e., prototype or template, is associated with each class or cluster. An MDC scheme classifies an unknown vector to one class or cluster if the distance between the unknown vector and the corresponding reference vector is minimum. This simple technique achieves fairly good performance in most cases. Since the computation of MDC is regular and simple, it can be performed by special VLSI arrays. A systolic array for the MDC which is capable of computing L_1, L_2 and L_∞ metrics is presented in this chapter.

Minimum-distance classification schemes can be easily extended to deal with strings. Now the bottleneck is the string-distance computation which is becoming popular due to the applications in, among others, syntactic pattern recognition, signal processing and information retrieval. We also present a systolic array for the MDC of strings.

This chapter is organized as follows. First we present a systolic array for the MDC of vectors. Next we describe a systolic array for string-distance computation. Finally seismic discrimination using these systolic arrays is shown as an application example.

4.1 Minimum-Distance Classification

Consider the classification problem of ℓ classes $C_1, C_2, \ldots, C_\ell$ where each pattern class C_j has a reference or template pattern M^j. An MDC scheme with respect to $M^1, M^2, \ldots, M^\ell$ classifies the unknown pattern X to class C_j if $d(X, M^j) = \min\{d(X, M^i), i = 1, 2, \ldots, \ell\}$, where $d(X, M^j)$ is the distance defined between X and M^j. In statistical pattern recognition X is a feature vector; in syntactic pattern recognition X is a structure such as string, tree or graph. For syntactic representation we shall discuss only the string case since it has wide application, and the tree pattern can always be converted to string by using traversal.

If $X = [x_1, x_2, \ldots, x_k]$ and $M^j = [m_1^j, m_2^j, \ldots, m_k^j]$ are vectors we can consider X and M^j as two points in feature space. Consequently

$$d_p(X, M^j) = (|x_1 - m_1^j|^p + |x_2 - m_2^j|^p + \ldots + |x_k - m_k^j|^p)^{1/p} \text{ and}$$

$$d_\infty(X, M^j) = \max(|x_1 - m_1^j|, |x_2 - m_2^j|, \ldots, |x_k - m_k^j|)$$

under L_p metric. When $p = 2$, $d_p(X, M^j)$ is the Euclidean distance. The minimum-distance classifier using the Euclidean distance is a linear classifier [4.1]. The time for classifying an unknown vector is proportional to $(\ell \times k)$, where ℓ is the number of reference vectors and k is the dimension of the feature vectors. This is not economical when the number of test samples is large, nor acceptable when real-time response is required. A systolic array, which will be discussed in the next section, can solve this problem.

The MDC is also extensively used in clustering, for example, K-means and ISODATA algorithms, where the unknown samples are assigned to the nearest clusters. Another application of MDC is to primitive recognition in wave form or shape analysis where pattern segments are converted into a finite number of primitives.

If X and M^j are strings, then $d(X, M^j)$ is the string-to-string distance. The computation of string-to-string distances varies depending on the applications. Two popular examples are the Levenshtein and time-warping distances. The Levenshtein distance between strings X and M^j, $d^L(X, M^j)$, is defined as the smallest number of transformations required to derive string M^j from X. The transformations include insertion, deletion and substitution of pattern primitives. The Levenshtein distance is a common choice in syntactic pattern recognition and information retrieval because of its error-correcting capability. The string-to-string distance can be used as a similarity measure when an exact match cannot be found.

The Levenshtein distance can be computed by the following dynamic programming algorithm. A simple example of computing the Levenshtein distance is given in Fig.4.1.

Algorithm 1. Computation of Levenshtein distance

  Input: Two strings $X = a_1 a_2 \ldots a_n$ and $Y = b_1 b_2 \ldots b_m$
  where a_i, b_j are pattern primitives for all $1 \leq i \leq n$, $1 \leq j \leq m$.

  Output: The Levenshtein distance $d^L(X,Y)$.

  Method:
  1) $\delta[0, 0] := 0$;
  2) For $i := 1$ to n do $\delta[i, 0] := \delta[i-1, 0] + 1$;
  3) For $j := 1$ to m do $\delta[0, j] := \delta[0, j-1] + 1$;

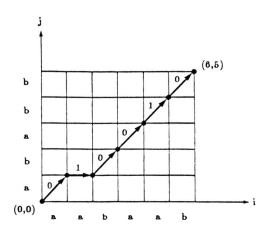

Fig.4.1. Transformation from string 'aabaab' to 'ababb'. The Levenshtein distance $d^L(aabaab, ababb) = 2$

4) For j:= 1 to m do
   For i:= 1 to n do
   $\delta[i, j] := \min\{\delta[i-1, j] + 1, \delta[i, j-1] + 1, \delta[i-1, j-1] + 1\}$;
5) $d^L(X,Y) := \delta[n, m]$.

The time complexity of Algorithm 1 is O(mn), where m and n are lengths of the two strings. Obviously this is the area where improvement of computation speed is needed. Dynamic programming algorithms are suitable for VLSI, particularly systolic array, implementations. We shall show the details in Sect.4.3.

Since the Levenshtein distance is not powerful enough, weights have been introduced associated with different transformations and primitives [4.2], resulting in a weighted Levenshtein distance. We use S(a, b) for the cost of substituting primitive **b** for **a**, D(a) for the cost of deleting **a** and I(a) for the cost of inserting **a**. For computation of a weighted Levenshtein distance we make the following changes to Algorithm 1: $\delta[i,0] := \delta[i-1, 0] + D(a_i)$ in Step (2), $\delta[0,j] := \delta[0,j-1] + I(b_j)$ in Step (3) and $\delta[i,j] := \min\{\delta[i,j-1] + I(b_j), \delta[i-1, j-1] + S(a_i,b_j), \delta[i-1,j] + D(a_i)\}$. The computation of weighted Levenshtein distance has the same time complexity O(mn) as Algorithm 1 with an additional weight table.

The time-warping distance between strings X and Y is defined as

$$d^{TW}(X,Y) = \sum_{k=1}^{K} d(c(k)), \quad \text{where}$$

$$d(c(k)) = d(i(k), j(k)) = \|a_i(k) - b_j(k)\| \quad ,$$

a_i, b_j are vectors and k is the index of a common time axis. This is used most often in speech recognition [4.3]. A wealth of variations in computation exist. A simple case may use

$$\delta[i,j] = \min\{\delta[i,j-1] + \|a_i - b_j\|, \delta[i-1,j-1] + \|a_i - b_j\|,$$
$$\delta[i-1,j] + \|a_i - b_j\|\} .$$

More complicated slope constraints, weight coefficients and symmetric constraints can be employed. This type of distance can also be applied to other signal-processing problems.

Virtually all string-to-string distance computations can use the same dynamic programming procedure as in Algorithm 1. They differ only in the computation of $\delta[i,j]$. Therefore we can use the same array structure but need a different structure for the processing unit and communication for different applications. In this chapter we intend to design VLSI minimum-distance classifiers to be used as special-purpose processors to enhance the power of the host computer.

4.2 Vector Distances

From the procedural point of view minimum-distance classification can be divided into two steps: distance computation and classification. Distance computation can be formulated as a pseudo-matrix-multiplication problem $[Y] = [M] \circ [X]$, i.e.,

$$\begin{bmatrix} y_1^1 & y_1^2 & \cdots & y_1^n \\ y_2^1 & y_2^2 & \cdots & y_2^n \\ \cdot & \cdot & & \cdot \\ \cdot & \cdot & & \cdot \\ y_\ell^1 & y_\ell^2 & & y_\ell^n \end{bmatrix} = \begin{bmatrix} m_1^1 & m_2^1 & \cdots & m_k^1 \\ m_1^2 & m_2^2 & \cdots & m_k^2 \\ \cdot & \cdot & & \cdot \\ \cdot & \cdot & & \cdot \\ m_1^\ell & m_2^\ell & \cdots & m_k^\ell \end{bmatrix} \circ \begin{bmatrix} x_1^1 & x_1^2 & \cdots & x_1^n \\ x_2^1 & x_2^2 & \cdots & x_2^n \\ \cdot & \cdot & & \cdot \\ \cdot & \cdot & & \cdot \\ x_k^1 & x_k^2 & \cdots & x_k^n \end{bmatrix}$$

$$= \begin{bmatrix} M^1 \\ M^2 \\ \cdot \\ \cdot \\ M^\ell \end{bmatrix} \circ [X^1 \ X^2 \ \ldots \ X^n] ,$$

where k, ℓ and n are the dimension of feature vectors, the number of classes and the number of input vectors, respectively; X^i, $1 \leq i \leq n$, is the ith input vector and M^j, $1 \leq j \leq \ell$, is the reference vector associated with the jth class. The distance between vectors X^i and M^j is $y^i_j = d(X^i, M^j)$, and $d(X^i, M^j)$ can be L_1, L_2 or L_∞ metrics. Classification is equivalent to a transformation of matrix [Y] into a vector of complex elements

$$Y' = [y^1, y^2, \ldots, y^n] \, , \quad \text{where}$$

$$y^i = (y^i_{q^i}, q^i), \quad y^i_{q^i} = \min(y^i_1, y^i_2, \ldots, y^i_\ell) \, .$$

Kung and *Leiserson* [4.4] proposed a hexagonal systolic array for the multiplication of two $n \times n$ band matrices. Since the matrices [M] and [X] in the present study are neither square nor banded and the dimension n of [X] may be very large, a hexagonal array is therefore impractical due to the large number of processors required. We use a square array for the pseudo-matrix-multiplication. This ℓ by k array is shown in Fig.4.2. The reference vectors enter from the bottom of the array and move up while the input vectors enter from the top and move down. To achieve parallel and pipelined processing the data elements of both input and reference vectors must be properly skewed as shown in Fig.4.2. The partial sums move from left to right. Since the input and reference data streams move in opposite directions, their data elements must be separated by one unit-time delay as shown in Fig.4.2, otherwise some data will be missing from the computation.

The input vectors are assumed to be fed in continuously. The reference vectors must also repeat their cycles, i.e., with the first reference vector coming right after the ℓth reference vector. After initiation the input vectors are delayed for $\ell-1$ unit times so that the first input vector and the first reference vector will meet at the first (top) row of the processor array. The sum, which is zero initially, will become distance on coming out at the end of each row. We call each processing element (PE) of the ℓ by k array a 'compute' processor. The functional diagram of the 'compute' processor is shown in Fig.4.3a, where x is a component of the input vector, u is a component of the reference vector and 'a' is the partial sum. Here we consider the L_1 metric, i.e., absolute-value metric, first.

The internal structure and data movement of the 'compute' processor are given in Fig.4.4a. Each 'compute' processor contains an arithmetic and logic unit (ALU) and four registers: A, B, U and X. The microoperations are as follows:

1a) Transfer data (serially) into register X from the upper PE.
1b) Transfer data (serially) into register U from the lower PE.

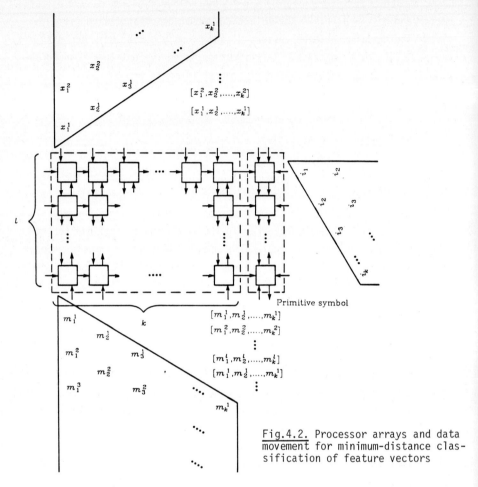

Fig.4.2. Processor arrays and data movement for minimum-distance classification of feature vectors

1c) Transfer partial sum (serially) into register A from the left PE.
2) $B \leftarrow X - U$.
3) $B \leftarrow |B|$.
4) $A \leftarrow A + B$.

Due to the communication overhead and pin limitation the data are transferred serially between PE's and memory. Therefore Step (1) takes 16 clock cycles to transfer one 16-bit word; Steps (2), (3) and (4) each take one clock cycle. The entire operation takes 19 clock cycles. The time for each PE to complete its entire operation, i.e., 19 machine cycles here, is called a unit time.

For L_2 metric, i.e., Euclidean distance, we need only to change the function of the 'compute' processor in Fig.4.3a from $b \leftarrow a + |x - u|$ to $b \leftarrow a + (x - u) \times (x - u)$. In microoperations we have to change Step (3) to $B \leftarrow B \times B$. The speed

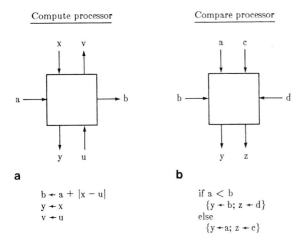

Fig.4.3a,b. Data flow and operations of each (a) 'compute' processor and (b) 'compare' processor

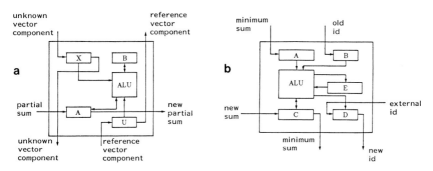

Fig.4.4a,b. Internal structure and register transfer of (a) 'compute' and (b) 'compare' processors

of multiplication depends on the implementation. A highly parallel multiplier can complete the operation in one clock cycle. Since taking square-root does not affect the minimum-distance selection, we can skip this computation. For L_∞ metric, i.e., maximum-value metric, we need to change the function of the 'compute' processor in Fig.4.3a from $b \leftarrow a + |x - u|$ to $b \leftarrow \max\{a, |x - u|\}$. In microoperations we have to change Step (4) to $A \leftarrow \max\{A, B\}$. This new step takes 2 clock cycles: one for maximum selection, the other for data transfer.

When the partial sum passes the kth 'compute' processor of the first row the output is the distance between vectors $[x_1^1, x_2^1, \ldots, x_k^1]$ and $[m_1^1, m_2^1, \ldots, m_k^1]$. It then enters the 'compare' processors, i.e., the rightmost column in Fig. 4.2, where the minimum distance will be selected. Since the data elements are separated by one unit time, the processors on alternate diagonals (from lower left corner to upper right corner) are idle. When vector $[x_1^1, x_2^1, \ldots, x_k^1]$ enters the third row it will meet vector $[m_1^3, m_2^3, \ldots, m_k^3]$; meanwhile, vector $[x_1^2, x_2^2, \ldots, x_k^2]$ will meet vector $[m_1^2, m_2^2, \ldots, m_k^2]$ at row one. Therefore X^1 meets reference vectors in the sequence $M^1, M^2, \ldots, M^\ell$; vector X^2 meets reference vectors in the sequence $M^2, M^3, \ldots, M^\ell, M^1$ and so forth. These operations are pipelined so that each PE is doing only part of the computation and passes the data and results to the next PE.

The functional diagram of the 'compare' processor is shown in Fig.4.3b, where **a** is the minimum distance computed so far with class identifier **c**; **b** is the distance just computed and **d** is the corresponding class identifier input externally. The internal structure and data movement are shown in Fig.4.4b. Each 'compare' processor contains an ALU, two 8-bit registers B, D and two 16-bit registers A, C. The microoperations are as follows.

1a) Transfer partial sum (serially) into register A from C of the upper PE.
1b) Transfer partial sum (serially) into register C from the left PE.
1c) Transfer class identifier (serially) into register B from D of the upper PE.
1c) Transfer class identifier (serially) into register D from external input.
2) $E \leftarrow A - C$.
3) If $a < c$ then $\{C \leftarrow A; D \leftarrow B\}$.

Step (1) takes 16 cycles to complete the transfers; Step (2) needs 1 cycle and Step (3) also needs 1 cycle. These three steps take 18 cycles, which is 1 cycle shorter than the 'compute' processor, therefore the 'compare' processor must be idle for one cycle in order to be synchronous with the 'compute' processor. The 'compare' processors compare the crurrent distance coming from the left with the minimum distance coming from above, and pass the smaller one to the lower processor. Class identifiers are fed in from the right in a similar format to those for data streams. The identifier streams should be delayed for $\ell + k - 1$ unit times so that the first identifier i_1 enters the first 'compare' processor at the same time as the distance between $[x_1^1, x_2^1, \ldots, x_k^1]$ and $[m_1^1, m_2^1, \ldots, m_k^1]$. In order to assign right identifier to right distance, the identifier streams must be arranged as shown in Fig.4.2.

With a uniprocessor, the MDC will take $O(\ell \times k)$ computations and $O(\ell - 1)$ comparisons to classify an input vector regardless of how many input vectors are involved. With the processor array of Fig.4.2 the classification result of the first input vector is available after $\ell \times k + 1$ unit times. However, a processor array is not designed for the processing of one single datum, instead, it is for a stream of data. In that case, a new result will come out every 2 unit times in Fig.4.2. Given ℓ reference vectors and feature vectors of dimension k, the array processor will take 2 unit times to get one result in steady state, while a uniprocessor takes $O(\ell \times k)$ time to complete the computation. The speedup is $\ell \times k/2$. In Fig.4.2, the results contain both the minimum distance and the class identifier, therefore no other processing is required.

4.3 String Distances

Nonnumeric computation has become more important and demanded more hardware algorithms, i.e., algorithms specially designed for hardware implementations, and also architectures recently, due to the increasing applications to artificial intelligence, database, information retrieval, language translation, pattern recognition, etc. One of the most important categories in nonnumeric computation is string pattern matching which is essentially string MDC. Character string matching is very important in information retrieval and dictionary lookup. The problems of string pattern matching can generally be classified into exact matching and approximate matching. For exact matching, a single string is matched against a set of strings, usually this particular string is embedded as a substring of the reference strings. *Foster* and *Kung* [4.5] designed a VLSI chip for exact pattern matching with wild card capability, where the input pattern enters from one end and the reference text enters from the other end of the linear array. By contrast, for approximate matching, we want to find a string from a finite set of strings which approximately matches the input string. Certainly we shall also find the string which exactly matches the input string if it does exist. A good survey of approximate string matching can be found in [4.6]. This section concentrates exclusively on approximate matching. Approximate string matching is based on the application of insertion, deletion and substitution to terminal symbols. An application example of approximate string matching which cannot be performed by exact string matching is the string clustering problems.

4.3.1 Levenshtein Distance

A portion of the dynamic programming diagram and its corresponding processor array is given in Fig.4.5. Each processor computes the partial sum

$$S_{i,j} = \min\{S_{i-j,j} + 1, S_{i-1,j-1} + S(a_i,b_j), S_{i,j-1} + 1\} \quad ,$$

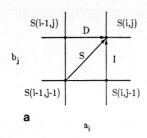 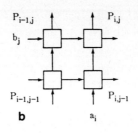

Fig.4.5. (a) Portions of dynamic programming diagram and (b) corresponding processor array

where $S(a_i,b_j) = 1$ if $a_i \neq b_j$; $S(a_i,b_j) = 0$ otherwise. The computation can be divided into three stages. The procedures are as follows.

Stage 1

1a) Transfer (serially) partial sum $S_{i-1,j-1}$ into D from the lower PE.
1b) Transfer (serially) primitive a_i into X from the lower PE.
1c) Transfer (serially) primitive b_j into Y from the left PE.
1d) Compare (serially) X with Y; output V = 0 if X = Y, V = 1 otherwise.
2) $D \leftarrow D + V$.

Stage 2

1a) Transfer (serially) partial sum $S_{i-1,j}$ into B from the left PE.
1b) Transfer (serially) partial sum $S_{i,j-1}$ into C from the lower PE.
1c) Send (serially) partial sum $S_{i-1,j}$ to D of the upper PE.
1d) Compare (serially) B with C, $A \leftarrow \min(B,C)$.
1e) Send (serially) contents of X to X of the upper PE.
1f) Send (serially) contents of Y to Y of the right PE.
2) $A \leftarrow A + 1$.
3) Compare (in parallel) A with D, $R \leftarrow \min(A,D)$.

Stage 3

1a) Send (serially) partial sum R to B of the left PE.
1b) Send (serially) partial sum R to C of the upper PE.

Stage 1 takes 17 clock cycles to complete the operation [16 for Step (1) and 1 for Step (2)]; Stage 2 takes 18 [16 for Step (1), 1 for Step (2) and 1 for Step (3)], and Stage 3 takes 16. Figure 4.6 shows the internal structure and the operations of processor element $P_{i,j}$ at Stages 1, 2 and 3. Each PE contains a set of registers, an ALU, a control unit and some other combinational logic. Registers A, B, C, D, V and R are general-purpose registers which are 16-bit long and connected to the ALU. Registers X and Y are 8-bit long, used

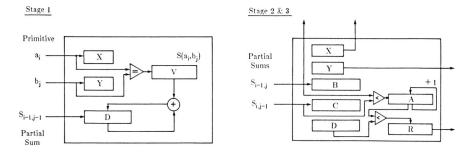

Fig.4.6. Internal structure and register transfer of PE P_{ij} at Stage 1, 2 and 3

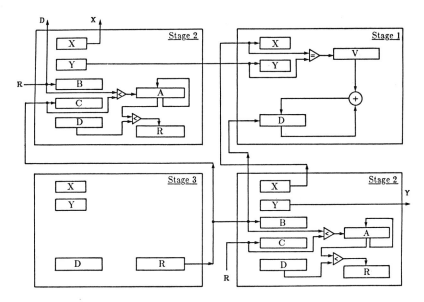

Fig.4.7. Data movement between PE's

to store primitives. This setup will be able to recognize character strings where each character is in ASCII code or the like.

Figure 4.7 shows the data movement between 4 neighboring PE's shown in Fig.4.5. All the processors at the same diagonal perform the same computation. This format will move one step forward every 18 clock cycles. Since each pair of strings needs only three diagonals at any time, the other processors can be used to compute distances between other strings. Therefore, data flow can be pipelined as shown in Fig.4.8. If we are matching an input string against a number of reference strings, the distance between the input string and the first reference string will emerge after $p \times 18$ clock

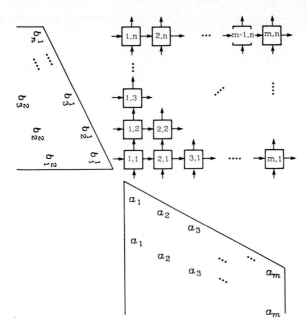

Fig.4.8. Processor array and data movement for computing string distance

cycles, where p is the number of diagonals in the array. After that, one string distance emerges every $3 \times 18 = 54$ clock cycles. Since Stage 1 and 3 have no conflict, they can be overlapped, i.e., one diagonal of the array can be used to perform Stage 3 of one string and Stage 1 of the next string at the same time, to increase the throughput.

The structure of processor array and data flow are shown in Fig.4.8. The reference strings enter from the left; the input string enters from the bottom. The same input string must be fed in continuously in order to compare with all the reference strings. Both input and reference strings must be properly skewed and separated as shown in Fig.4.5, so that they will arrive at the right processors at the right time. The bookkeeping and selection of minimum distance can be done by a special-purpose processor or the host computer. It simply selects $\min\{d(X,R^i), d_{min}(i)\}$, where $d_{min}(i) = \min\{d(X,R^j), 1 \leq j < i\}$; X is the input string and R^i is the ith reference string. One practical problem concerns the dimensions of the processor array. The number of rows can be set to the maximum length of the reference strings. Since the length of the input string is unknown, the number of columns can be set arbitrarily. If an input string exceeds the array size, it should be handled by the host computer or preprocessor, because the interruption of the regular computation pattern in a VLSI array will greatly reduce its efficiency. This situation can be kept to a minimum by selecting a reasonably large array size. A shorter string will be padded with blank to made it equal to the array dimension.

Suppose both input and reference strings have length ℓ. With a uniprocessor the matching process for the unknown string will take $O(\ell \times \ell)$ unit operations. With the array processor, it takes only 3 unit times.

4.3.2 Weighted Levenshtein Distance

Since a weighted Levenshtein distance is usually more suitable to practical application, we now propose a VLSI architecture for its computation. The major problem here is to store the weight table in each processor, which must be easy to implement and fast for access. Therefore, the programmable logic array (PLA) is a natural selection [4.7]. It is a special type of read-only memory, and easily implemented in a VLSI system. Figure 4.9 shows the PLA implementation of a weight table for a real application discussed in Sect.4.4, which includes insertion, deletion and substitution. Thirteen pattern primitives labeled 'a' - 'm' were selected. Assume that the primitives are represented by ASCII code. We take the 4 least significant bits from the primitives for our internal computation, i.e., 'a' = 0001, 'b' = 0010, etc. An input register is connected to the AND plane and an output register is connected to the OR plane. Both registers are 8-bit long. Registers X and Y contain primitives which form the entries of the weight table. Input pair (X = a, Y = b) represents the substitution of 'b' for 'a'. Input pair (X = a, Y = 0000) represents the deletion of 'a' and (X = 0000, Y = a) means the insertion of 'a'. It takes only two clock cycles, one to input register, the other to output register, to access the PLA.

Except for the weight table the computation procedure is similar to the previous one. The internal structure of the PE's is given in Fig.4.10. Each PE has an ALU, a PLA (with registers Q and S), a control unit, two 8-bit registers X, Y, and three 16-bit registers B, C and D. Register Z contains constant zero as null string. The data movement is similar to that in Fig. 4.7.

Stage 1

1a) Transfer (serially) partial sum $S_{i-1,j-1}$ into D from the lower PE.
1b) Transfer (serially) primitive a_i into X from the lower PE.
1c) Transfer (serially) primitive b_j into Y from the left PE.
2) Load (parallel) X and Y into Q, output $S(a_i, b_j)$ in S.
3) Compute $D \leftarrow D + S$.
4) Load (parallel) X and Z into Q out $D(a_i)$ in S.

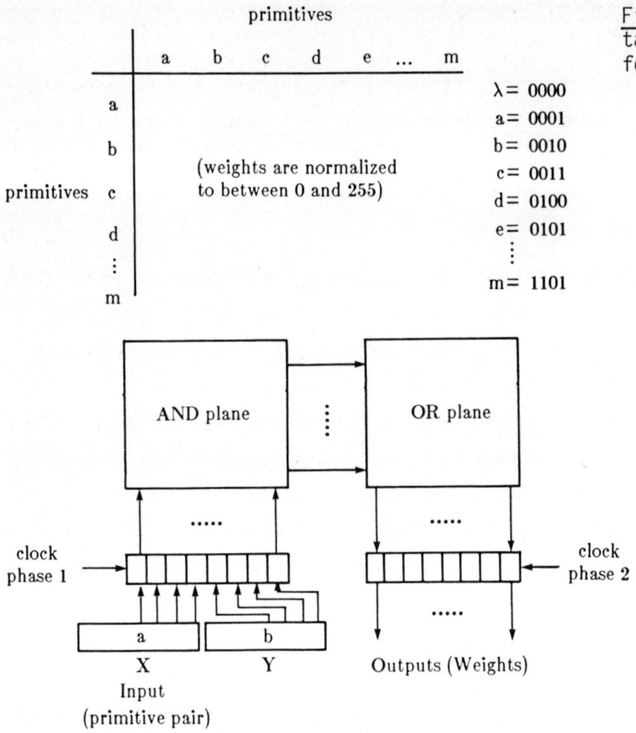

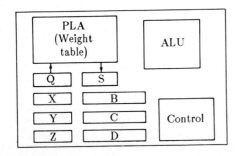

Fig.4.9. PLA implementation of weight table for seismic recognition

Fig.4.10. Internal structure of the PE for weighted string distance computation

Stage 2

1a) Transfer (serially) partial sum $S_{i-1,j}$ into B from the left PE.
1b) Transfer (serially) partial sum $S_{i,j-1}$ into C from the lower PE.
1c) Send (serially) partial sum $S_{i-1,j}$ to D of the upper PE.
1d) Send (serially) contents of X to X of the upper PE.
1e) Send (serially) contents of Y to Y of the right PE.
2a) Compute $B \leftarrow B + S$.
2b) Load (in parallel) Z and Y into Q, output $I(b_j)$ in S.

3a) Compute C ← C + S.
3b) Compute B ← min(B,D).
4) Compute D ← min(B,C).

Stage 3

1a) Send (serially) partial sum in D to B of the right PE.
1b) Send (serially) partial sum in D to C of the upper PE.

In Stage 1, Step (1) takes 16 cycles [(a), (b) and (c) operate in parallel], Step (2) takes 3 cycles (1 for loading, 2 for PLA reading), Step (3) takes 1 cycle and Step (4) takes 3 cycles [same as Step (2)]. In Stage 2, Step (1) takes 16 cycles, Step (2) takes 3 Cycles [(a), (b) operate in parallel], Step (3) takes 2 cycles and Step (4) takes 2 cycles. Stage 3 takes 16 cycles [(a) and (b) both take 16 cycles and can be executed in parallel]. Therefore Stage 1 takes 23 cycles, Stage 2 take 23 cycles, and Stage 3 takes 16 cycles. As usual, Stage 3 can be overlapped with Stage 1 to save processing time. Due to the weight computation, this system takes longer than the previous one.

4.3.3 Time-Warping Distance

A VLSI array for pattern matching by dynamic time-warping has been proposed in [4.8]. The array structure is similar to that of Fig.4.8. Each processing element computes the partial sum

$$S_{i,j} = D_{i,j} + \min\{S_{i-1,j}, S_{i-1,j-1}, S_{i,j-1}\} ,$$

where $D_{i,j} = |x_i - y_j|$ is the difference measure between the ith element of string X and the jth element of string Y, and x_i and y_j can be either scalars or feature vectors. If x_i and y_j are vectors the computation is more complicated. The whole vectors must be sent in serially, i.e., word serial and bit serial, due to communication overhead and pin limitation. The computation of $D_{i,j}$ is executed sequentially, one vector element at a time. Consequently, the throughput for vectors is slower than that for scalars. The vector elements can be held in one processing element until the computation of $D_{i,j}$ is completed, or they can be passed immediately to the next PE's. Dynamic time-warping is usually used in speech recognition where string elements are vectors. A practical selection of vector dimension may range from 5 to 15. Only VLSI processor arrays can make real-time applications possible.

4.4 Examples of Application

We have applied by simulation the proposed VLSI arrays to real seismic discrimination problems [4.9]. This intended to serve two purposes. First, we showed how to apply these VLSI arrays to real problems. Second, we wanted to verify the design of these VLSI arrays and to evaluate their performance. We had some 300 real seismic records and wanted to classify them into two classes—earthquake and nuclear explosion. We used a syntactic appraoch in [4.9] where the seismic signal is first converted into a string of pattern primitives and then the nearest-neighbor decision rule is applied for classification. A VLSI array was designed for feature extraction, which computes the number of zero crossings and total energy of the seismic signal. This converts each pattern segment into a feature vector. The primitive recognition procedure is essentially a vector classification problem. A reference vector for each pattern primitive was determined during the training stage. A VLSI minimum-distance classifier, as shown in Sect.4.2, assigned each feature vector a pattern primitive. Nearest-neighbor decision procedure can be carried out by a minimum string-distance classifier as shown in Sect.4.3.2. The same data have been tested by sequential algorithms and simulated VLSI arrays. The classification results and all the intermediate results are the same in both cases, verifying the correctness in the function level of our VLSI array design.

The computation time in our simulation is shown in Table 4.1. The computation time using the sequential algorithm is also given for comparison. Note that the listed computation time is approximate and average time which should be used for comparison only. Suppose that we are dealing with a large amount of data. Similar to the definition of speedup for multioperation computers, we define the estimated speedup (ES) of a (systolic) processor array as

$$ES = \frac{\text{time interval between consecutive results using a sequential computer}}{\text{time interval between consecutive results using a processor array}}.$$

Therefore the ES for primitive recognition is $39/2 = 19.5$ and for string matching is $20/3 = 6.67$. The numerators are the numbers of operations for getting one result using the sequential algorithm, and the denominators are the time intervals between consecutive results for VLSI arrays as shown in the previous sections. Note that the ES for string matching in our experiment is $20/3 = 6.67$ instead of $20 \times 20/3 = 133$. This is because we consider only substitution errors in [4.9], therefore the number of operations is

Table 4.1. Computation time of sequential algorithm, simulated computation time for VLSI arrays using sequential computer, real speedups, estimated speedups and speedup ratio

	Sequential algorithm [s]	Simulated VLSI arrays [ms]	Real speedup	Estimated speedup	Speedup ratio
Feature vector classification	0.005	0.3	17.4	19.5	89%
String classification	0.07	15	4.7	6.7	70%

proportional to string length, i.e., 20. If insertion and deletion errors are to be considered, then the whole dynamic programming matrix as shown in Fig.4.1 should be considered. In the seismic discrimination problem the size of the matrix is 20×20. The real speedup (17.4) for primitive recognition is slightly less than the ES (19.5), which is about 89% of the ES, due to the complexity of array structure and data flow. Some time is spent on data movement. The real speedup (4.67) for string matching is even less than the ES (6.67), which is only 70% of the ES, because the array structure and data flow are more complicated.

The simulations were performed on a VAX-11/780 computer. To compare simulation with sequential algorithm results we use the same high-level languages (C, FORTRAN and PASCAL). Therefore, there are many overheads in language translation and program execution, which is one of the reasons for lower speedup. Another reason is the data movement which can be performed in parallel within the VLSI arrays, but cannot be fully simulated in a sequential computer. Since most systolic arrays are hardwired, i.e., nonprogrammable, there is no instruction decoding or memory fetch for each instruction. The real computation speeds of the proposed VLSI arrays when fabricated should stay close to the analytical results as shown in the previous sections, i.e., 2 unit times for primitive recognition and 3 unit times for string matching using weighted Levenshtein distance.

4.5 Summary

Minimum-distance classification is a common technique in both vector and string applications. Vector-distance computation has been studied for quite a long time. On the other hand, string-distance computation is relatively new, but is getting popular. Vector distance computation and comparison is formulated by pseudo-matrix-multiplication. A VLSI square array is designed to process continuous input vectors. Classification results are available every two unit times. With minor changes in microoperations it can compute L_1, L_2 and L_∞ metrics. We can find applications of this VLSI minimum-distance classifier in statistical pattern recognition, clustering and primitive recognition in wave-form and shape analysis.

String-distance computation is becoming important due to the application in artificial intelligence, database and information storage and retrieval, and syntactic pattern recognition. The principal technique for string-distance computation is dynamic programming where the optimal match is computed between the two strings. The matching between two strings is handled in terms of transformation from one string to the other, involving insertion, deletion and substitution of pattern primitives. A square array is shown for fast computation of string-to-string distance. Each processing element computes the partial distance based on the available information. This computation can be pipelined so that string distance is available every two or three unit times. Weighted string distance needs a weight table. This can be easily implemented using a programmable logic array. In some cases, like speech recognition, the string elements are vectors and only substitution is considered. Then the array structure is unchanged whereas data transfer and microoperation sequence need some modifications.

We have applied these VLSI minimum-distance classifiers by simulation to real seismic classification problems for design verfication and performance evaluation. The functional simulations showed that the design is correct but the real speedup does not match the estimated speedup. The lower speedup is because we simulated only the function and the data transfer is not executed concurrently.

References

4.1 K.S. Fu: *Syntactic Pattern Recognition and Applications* (Prentice-Hall, Englewood Cliffs, NJ 1982)
4.2 K.S. Fu, S.Y. Lu: A Clustering Procedure for Syntactic Pattern Recognition, IEEE Trans. SMC-**7**, 734-742 (1977)
4.3 H. Sakoe, S. Chiba: Dynamic Programming Algorithm Optimization for Spoken Word Recognition, IEEE Trans. ASSP-**26**, 43-49 (1978)
4.4 H.T. Kung, C.E. Leiserson: Systolic Arrays (for VLSI), in C.A. Mead, L.A. Conway: *Introduction to VLSI Systems* (Addison-Wesley, Reading, MA 1980) Sect.8.3
4.5 M.J. Foster, H.T. Kung: The Design of Special-Purpose VLSI Chips, Computer **13**, 26-40 (1980)
4.6 P.A.V. Hall, G.R. Dowling: Approximate String Matching, ACM Comput. Surveys **12**, 381-402 (1980)

4.7 C.A. Mead, L.A. Conway: *Introduction to VLSI Systems* (Addison-Wesley, Reading, MA 1980)
4.8 B. Ackland, N. Weste, D.J. Burr: An Integrated Multiprocessing Array for Time Warping Pattern Matching, Proc. 8th Ann. Symp. on Comput. Arch., Minneapolis (1981) pp.197-215
4.9 H.H. Liu, K.S. Fu: A Syntactic Approach and VLSI Architectures for Seismic Signal Classification, Tech. Rep., TR-EE 83-5, School of Electrical Engineering, Purdue University, West Lafayette, IN (1983)

5. Design of a Pattern Cluster Using Two-Level Pipelined Systolic Array

L.M. Ni and A.K. Jain

Cluster analysis is a valuable tool in exploratory pattern analysis especially when very little prior information about the data is available. In unsupervised pattern recognition and image segmentation applications, clustering techniques play an important role. The squared-error clustering technique is the most popular clustering technique. Due to the iterative nature of squared-error clustering, it demands substantial CPU time even for modest numbers of patterns. Recent advances in VLSI microelectronic technology triggered the idea of implementing squared-error clustering directly in hardware. A two-level pipelined systolic pattern clustering array is proposed in this chapter. The memory storage and access schemes are designed to enable a rhythmic data flow between processing units. Each processing unit is pipelined to enhance further the system performance. The total processing time for each pass of pattern labeling and cluster center updating is essentially dominated by the time required to fetch the pattern matrix once. Detailed architectural configuration and the system performance evaluation are presented. The modularity and the regularity of the system architecture make it suitable for VLSI implementation.

5.1 Background

Cluster analysis is a valuable tool in exploratory pattern analysis whose goal is to find natural groupings in a given data set [5.1]. The clusters obtained should have the property that patterns belonging to the same cluster are more similar to each other than patterns belonging to different clusters. The output of a clustering algorithm helps the user form hypotheses about the structure of the data.

Over the past twenty years virtually every scientific discipline including social science, biology, psychology, taxonomy, psychiatry, statistics, computer science, and engineering have utilized clustering methodology to analyze and interpret experimental data. Cluster analysis is especially useful when very little prior information about the data is available. This is the reason why clustering techniques play an important role in unsupervised pattern recognition and image segmentation. Specific problem areas utilizing clustering techniques include syntactic pattern recognition [5.2], speaker recognition [5.3], image segmentation [5.4], classification of multispectral scanner data [5.5], and image registration [5.6].

These numerous applications of clustering have resulted in an enormous collection of clustering algorithms [5.7]. Most clustering algorithms are based on the following three clustering techniques: minimizing *squared-error clustering*, *hierarchical clustering*, and *graph-theoretic clustering* [5.8]. Because squared-error clustering algorithms operate directly on the pattern matrix, they are the most popular in pattern recognition and image processing (PRIP) applications. The most well-known squared-error clustering algorithm is the ISODATA [5.8]. Due to the iterative nature of the squared-error clustering algorithms, they demand substantial CPU time even for a modest number ($\doteq 200$) of patterns.

During the past decade, considerable efforts have been devoted to developing special computer architectures for PRIP [5.9]. Recent advances in VLSI microelectronic technology have triggered the idea of implementing PRIP algorithms directly in specialized hardware chips. Many attempts have been made to develop special VLSI devices for such purposes. Some candidate PRIP algorithms that might be suitable for considering VLSI implementation are indicated in [5.10]. Squared-error pattern clustering involves a large amount of matrix or vector computations. The systolic architecture is thus especially suited for such applications [5.11].

In systolic architectures, the step processor consisting of a multiplier and an adder forms the basis for many matrix manipulation networks [5.12,13]. It has been shown that by pipelining the step processor or other building cells, the system throughput can be greatly increased [5.14] and the number of building cells can be significantly reduced [5.15]. However, feedback paths are usually involved in such two-level pipelined systolic arrays [5.15]. Therefore, the scheduling and the synchronization of data flow in the systolic array are more complicated and must be carefully designed.

After briefly describing the squared-error pattern clustering techniques in Sect.5.2, the systolic pattern clustering array is detailed in Sect.5.3. Five different pipelined processing units are used in our design. Both the system architecture and the memory storage and access schemes are considered. In Sect.5.4, we examine the system operating characteristics. Two major processes, label reassignment and cluster center updating, are performed in an overlapping fashion. The system performance evaluation is also included in Sect. 5.4.

5.2 Description of Squared-Error Pattern Clustering

Clustering techniques that minimize a squared error criterion have been extensively used in pattern recognition [5.8]. The central idea of this technique is to choose some initial partition of the patterns and then alter cluster memberships so as to obtain a better partition [5.16]. Various methods of implementing squared-error clustering technique have been studied in the past and comparatively surveyed [5.8,16]. Among them, Forgy's method, the ISODATA method, the k-means method, and the hill-climbing method are most frequently used. These methods differ only in the heuristic tactics employed. In this study, we do not intend to implement any particular clustering method. The heuristic part is left out in our design so that the user may specify any method or even one's own method. Only the central part of squared-error clustering technique, which is the most time-consuming part, is implemented in the form of systolic architecture. Also, the systolic architecture will provide useful information for adjusting the heuristic tactics. In what follows, we shall first briefly describe the squared-error clustering technique and its characteristics.

Let N represent the number of patterns (N vectors), which form a data set, to be partitioned into clusters. Each pattern has M features. Usually, M is far less than N and is in the range of 1 to 30. Let P be an $N \times M$ *pattern matrix* where the element $p(i,j)$ is a floating-point scalar representing the jth feature of the ith pattern for $0 \leq i \leq N-1$ and $0 \leq j \leq M-1$ and $P(i)$ is a $1 \times M$ vector representing the ith pattern. A clustering is a partition [S(0), S(1),...,S(K-1)] of the integers [0, 1,..., N-1] that assigns each pattern a single cluster label, where K, the number of distinct clusters, is data dependent and is usually less then 20. Let $L(i)$ represent the label of the ith pattern. Obviously, $0 \leq L(i) \leq K-1$ for $0 \leq i \leq N-1$. Formally, the kth cluster can be expressed as

$$S(k) = \{i | L(i) = k \text{ and } 0 \leq i \leq N-1\} \quad \text{for } 0 \leq k \leq K-1 \ . \tag{5.1}$$

For each cluster S(k), its cluster center C(k) is a $1 \times M$ vector where each element $c(k,j)$ is defined as

$$c(k,j) = \frac{1}{|S(k)|} \sum_{i \in S(k)} p(i,j) \quad \text{for } 0 \leq k \leq K-1 \text{ and } 0 \leq j \leq M-1 \ . \tag{5.2}$$

A K×M *cluster center matrix*, C, can be defined based on the elements specified in (5.2). The squared Euclidean distance d^2 between any two vectors, $A=(a_j)$ and $B=(b_j)$, is defined as

$$d^2(A,B) = \sum_{j=0}^{M-1} (a_j - b_j)^2 \quad . \tag{5.3}$$

The squared error for the kth cluster is defined as

$$e^2(k) = \sum_{i \in S(k)} d^2(P(i),C(k)) \quad \text{for } 0 \leq k \leq K-1 \quad . \tag{5.4}$$

and the squared error for the clustering is

$$E^2(K) = \sum_{K=0}^{K-1} e^2(k) \quad . \tag{5.5}$$

The objectives of a squared-error clustering technique are to define, for a given K, a clustering that minimizes $E^2(K)$ and to find a suitable K, much smaller than N. The problem of finding a suitable value of K for given data is difficult and is an important topic in clustering validity studies [5.1]. In practice, the clustering algorithm is applied repeatedly for different values of K and the resulting partitions are evaluated to obtain the "best" partition.

One major task of the squared-error clustering technique, called *label reassignment process*, is to associate each pattern P(i) with a suitable label L(i). The nearest centroid clustering assignment is the most frequently employed. From a given set of cluster centers, distances between each pattern and each cluster center are evaluated according to (5.3). For the ith pattern, its new label will be

$$L(i) = k \text{ where } d(P(i),C(k)) = \min\{d(P(i),C(k')), \text{ for } 0 \leq k' \leq K-1\} \quad . \tag{5.6}$$

Another major task, called the *cluster center updating process*, is to update the coordinates of each cluster center if the pattern label assignment has changed the membership of clusters as indicated in (5.2). Two approaches have been used in determining when the cluster centers will be updated. The k-means method invokes the updating process after each pattern assignment which changes the membership. Thus, the cluster centers may be updated after each label assignment. For most of the other methods, the cluster centers will remain fixed for a complete pass during the reassignment process for the entire set of patterns. The k-means method is very inefficient for parallel implentation due to its inherent sequential property. For

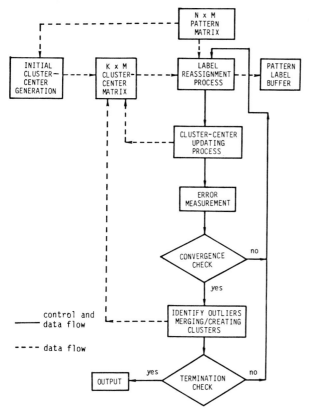

Fig.5.1. Flow-chart description of the iterative squared-error clustering technique

hardware design, it implies that most of the hardware resources will be wasted and cannot be fully utilized. In our design, for this reason, the cluster center updating process will be invoked only after the completion of label assignment process on the entire data set.

The functional description of the squared-error clustering technique is depicted in Fig.5.1. The solid lines indicate the direction of control and data flow and the dashed lines indicate the direction of data flow only. The initial cluster center generation box is a heuristic procedure to be supplied by the user. A variety of initial cluster-center generation methods have been employed [5.16]. Other major heuristic procedures, which have been extensively studied, are the criteria for identifying the outliers (patterns sufficiently far from cluster centers), for creating new clusters, and for deleting and merging clusters [5.8]. In order not to fix any particular heuristic in hardware, these procedures are not included in our design and are programmable subject to user-imposed restrictions. However, our design will provide the information to help the user in implementing the heuristic procedures.

The main body of the clustering technique repeats the label reassignment process and the cluster-center updating process until either convergence is achieved (no changes in cluster labels) or the number of passes reaches a predefined threshold. It is not possible to say a priori how many repetitions will be required to achieve convergence in any particular problem. However, empirical evidence indicates that five repetitions or less will ordinarily suffice; only infrequently will more than ten repetitions be needed [5.16]. For these special cases, a predefined threshold for the number of passes is thus needed.

5.3 The Systolic Pattern Clustering Array

As indicated in the previous section, our efforts will center on the hardware design of the label reassignment process and the cluster-center updating process as well as the memory design of storage and access schemes for the pattern matrix, cluster-center matrix, and some other buffers. Our objective is to design a special-purpose architecture such that the processing speed or the computation on each pattern can match the memory speed or the supply of each pattern. A systolic array architecture is best suited to achieve this goal [5.11].

The functional organization of our design is shown in Fig.5.2. Each pass involves a label reassignment process, a cluster-center updating process, and the convergence check. The convergence check is done by recording the number of patterns that have changed their membership since the last pass. If the total number of changes is greater than a user-defined threshold, another pass will be initiated. In the following subsections, we shall describe in detail the systolic architecture for pattern clustering.

5.3.1 Pipelined Processing Units

Since each element in the pattern matrix is a floating-point number, pipelined design of arithmetic operators can efficiently process a large volume of floating-point data or vector data [5.17]. Vector operations can be classified into four primitive types: $f_1: V \rightarrow V$, $f_2: V \rightarrow S$, $f_3: V \times V \rightarrow V$, and $f_4: V \times S \rightarrow V$ as shown in Fig.5.3, where V and S denote vector and scalar operands, respectively [5.17]. For example, vector square rooting is an f_1 operation, vector summation is an f_2 operation, vector addition is an f_3 operation, and the scalar-vector product is an f_4 operation. The combination of these primitive vector operations can perform a variety of vector opera-

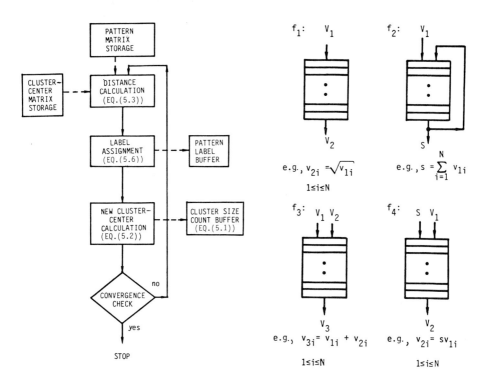

Fig.5.2. Functional block diagram of the proposed squared-error clustering implemented in hardware

Fig.5.3. Four basic primitive pipeline organizations and the corresponding examples

tions. For example, the dot product of two vectors is generated by applying vector multiplication (type f_3) and then vector summation (type f_2).

The f_2-type operation is also called a *vector reduction* operation because it accepts a vector input and generates a scalar output [5.18]. The vector reduction operation is performed by a recursive approach because of the feedback path involved. In the recursive evaluation, each output of the pipeline depends on the previous outputs. Improper or inefficient control of such feedback around a pipeline can destroy the efficiency and decrease its throughput. Two scheduling methods, *symmetric vector reduction* and *asymmetric vector reduction*, have been proposed to evaluate efficiently a single vector reduction [5.18].

For matrix manipulations, each matrix consists of many vectors such as rows or columns. The pipeline utilization can be further improved by allowing multiple vector reductions. For example, given a $G \times M$ matrix, we want to evaluate the row sums of the matrix in a pipeline adder with G segments. It is essentially equivalent to performing the vector summation of G vectors

TIME (PIPELINE CYCLES)

	1	2	3	4	5	6	7	8	9	10	11	12	13	14	15
G1	1,1	2,1	3,1	4,1	1,2	2,2	3,2	4,2	1,3	2,3	3,3	4,3			
G2		1,1	2,1	3,1	4,1	1,2	2,2	3,2	4,2	1,3	2,3	3,3	4,3		
G3			1,1	2,1	3,1	4,1	1,2	2,2	3,2	4,2	1,3	2,3	3,3	4,3	
G4				1,1	2,1	3,1	4,1	1,2	2,2	3,2	4,2	1,3	2,3	3,3	4,3

a →S(1) →S(2) →S(3) →S(4) (at cycles 12, 13, 14, 15)

$S(i) = a(i,1) + a(i,2) + a(i,3)$ for $i=1,\ldots,4$ (G=4, M=3)

The pipeline adder has 4 segments: G1, G2, G3, and G4.
The symbol i,j indicates that the corresponding segment is operating on elements $s(i,j-1)$ and $a(i,j)$, where $s(i,j-1)=a(i,1)+\ldots+a(i,j-1)$ for $i=1,\ldots,4$ and $j=1,\ldots,3$.

TIME (PIPELINE CYCLES)

	1	2	3	4	5	6	7	8	9	10	11	12	13	14	15	16	17	18	19	20
G1	1,1	2,1	3,1	1,2	2,2	3,2	1,3	2,3	3,3	4,1	5,1	6,1	4,2	5,2	6,2	4,3	5,3	6,3		
G2		1,1	2,1	3,1	1,2	2,2	3,2	1,3	2,3	3,3	4,1	5,1	6,1	4,2	5,2	6,2	4,3	5,3	6,3	
G3			1,1	2,1	3,1	1,2	2,2	3,2	1,3	2,3	3,3	4,1	5,1	6,1	4,2	5,2	6,2	4,3	5,3	6,3

b →S(1) →S(2) →S(3) (at cycles 10, 11, 12)   →S(4) →S(5) →S(6) (at cycles 18, 19, 20)

$S(i) = a(i,1) + a(i,2) + a(i,3)$ for $i=1,\ldots,6$ (G=3, M=3, B=2)..

The pipeline adder has 3 segments: G1, G2, and G3..

<u>Fig.5.4a,b.</u> Examples of interleaved multiple vector reduction: (a) the 4 by 3 matrix is not partitioned; (b) the 6 by 3 matrix is partitioned into two 3 by 3 submatrices

with M elements in each vector. If the matrix is supplied one element per cycle in column-major order and each pipeline segment takes one cycle delay, the first scalar row sum will come out after GM cycles. In the subsequent G-1 cycles, the remaining G-1 results will come out. In total, it takes G-1+GM cycles to get the final result of G scalar row sums. Figure 5.4a demonstrates an example of a 4×3 matrix evaluated in a 4-segment pipeline adder. Note that it takes GM cycles to fetch all the elements from the memory. This scheduling method is also called *interleaved multiple vector reduction*. If the matrix size is (BG) × M, the matrix can be partitioned into B G × M submatrices. In this case, draining out the pipeline in processing one submatrix can be performed simultaneously with filling up the pipeline for processing the next submatrix. The total processing time will be (G-1)+BGM cycles. In other words, if the pipeline is filled up, it can generate G scalar results per GM cycles. Figure 5.4b demonstrates an example for a 6×3 matrix evaluated in a 3-segment pipeline adder. Details of interleaved multiple vector reduction can be found in [5.15].

5.3.2 An Overview of System Architecture

A two-level pipelined architecture for pattern clustering is shown in Fig. 5.5 and will be explained in detail in the subsequent sections. Basically, there are two major processing parts. The upper part including the DSA (Difference-Square-Accumulate) cells and the pipeline comparator is for the purpose of label reassignment process. The lower part is dedicated to the cluster-center updating. The pattern matrix storage P provides two output streams, one for the label reassignment (column major) and the other for the cluster-center updating (row major). The K cluster centers are stored in K memory modules, $C(0), C(1),\ldots,C(K-1)$. A *label buffer* stores the new label for each pattern, and K *cluster counters* are needed to record the number

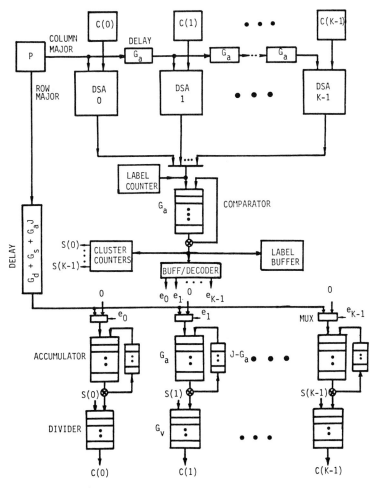

Fig.5.5. The systolic architecture for squared-error pattern clustering

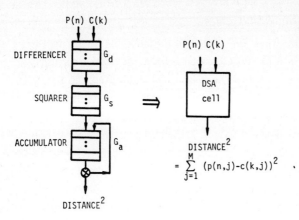

Fig.5.6. The pipeline organization and the logic symbol of the DSA cell

of patterns in each cluster. A few delays (shift registers) are used to synchronize the data flow. If the capacity of memory modules and buffers is large enough, the only restriction to this design is the maximum number of cluster centers K. Usually, the value of K is less than 20. With a minor modification, two such designs can be cascaded in parallel to accommodate a greater value of K.

In our design, five different pipeline processing units are used as basic building elements. A pipeline *differencer* with G_d segments is an f_3 operation which performs the vector subtraction. A pipeline *squarer* with G_s segments is an f_1 operation which performs the vector squaring. A pipeline *accumulator* with G_a segments is an f_2 operation which performs the vector summation. The above three segments are combined into one cell, namely the *DSA cell*, as shown in Fig.5.6. The function of the DSA cell is to calculate the squared Euclidean distance between a pattern vector and a cluster-center vector. A pipeline *comparator* with G_a segments is an f_2 operation which finds the index of the element with the minimum value in a vector. We assume that both the comparator and the accumulator have the same number of segments. In actual implementation, dummy segments may be inserted to make them equal. Finally, a pipeline *divider* with G_v segments is used to perform vector-scalar division, an f_4 operation.

In each pass, the new cluster centers will be evaluated and can be collected at the bottom of the design and saved in the cluster-center memory modules. The label buffer will also record the information on the total number of patterns that have changed their membership. Based on this information, another pass of the evaluation procedure may be triggered.

5.3.3 Memory Storage and Access Schemes

The pattern matrix P(N × M) and the cluster-center matrix C(K × M) must be stored in a way such that they can provide continuous input streams to the systolic architecture. The N by M pattern matrix is partitioned into a number of disjoint blocks. Each block is a G_a by M submatrix. There are $\lceil N/G_a \rceil$ blocks and the ith pattern belongs to the $\lfloor i/G_a \rfloor^{th}$ block. A physical pattern matrix, N by M, is stored in two independent memory modules, PM0 and PM1, in an interleaved fashion as shown in Fig.5.7. A virtual pattern matrix, N by J, is indicated by the dashed line, where $J = \max\{M, K, G_a\}$.

The virtual pattern matrix is needed to synchronize the system data flow. This fact will be explained in a later section. The pattern blocks are fetched according to their indices in the ascending order. Each block can be fetched either in the row-major or the column-major access scheme as illustrated in Fig.5.8. The row-major access scheme is used for evaluating the cluster centers and the column-major access scheme is used for labeling the patterns. To supply two output pattern streams simultaneously, two independent memory modules are needed. The ith block accessed in column major is

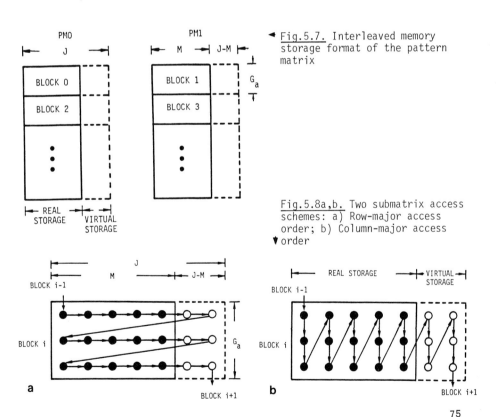

Fig.5.7. Interleaved memory storage format of the pattern matrix

Fig.5.8a,b. Two submatrix access schemes: a) Row-major access order; b) Column-major access order

fetched concurrently with the row-major access of the $(i-1)^{th}$ block. In other words, each block will be fetched twice in two consecutive turns, first in the column-major scheme, followed by the row-major scheme. Each turn takes $G_a J$ cycles. More precisely, at the tth cycle two elements, $p(i,j)$ and $p(i',j')$, are fetched, where each pipeline segment takes one cycle and each memory module can supply one element per cycle and

$$\begin{cases} i = G_a \lfloor \frac{t-1}{G_a J} \rfloor + \lfloor \frac{(t-1) \bmod G_a J}{J} \rfloor & 1 \leq t \leq NJ \\ j = \lfloor \frac{(t-1) \bmod G_a J}{G_a} \rfloor \end{cases} \quad (5.7)$$

$$\begin{cases} i' = \lfloor \frac{t - G_a J - 1}{J} \rfloor \\ j' = (t - G_a J - 1) \bmod J \end{cases} \quad (G_a J \leq t \leq (NJ + G_a J)) \quad (5.8)$$

Note that $p(i,j)$ is located in block $\lfloor i/G_a \rfloor$ and $p(i',j')$ is located in block $\lfloor i'/G_a \rfloor = \lfloor i/G_a \rfloor - 1$. The fetch of a nonexistent location, i.e., $M < j, j' \leq J$, implies no physical memory access. Instead, a dummy input is assumed.

The K cluster centers are stored in K independent memory modules as shown in Fig.5.9. Each memory module should store up to J elements. A linear skewed storage scheme is used with the skewing distance equal to 1 [5.19]. In such a storage scheme, $c(k,j)$ is stored in memory module k at the location $(k+j) \bmod J$. All memory modules are fetched simultaneously at the same location to supply input to the DSA cells. Cluster centers will be fetched every G_a cycles and the data fetched will be used for G_a consecutive cycles. At the tth cycle, data stored in location $[(t-1 \bmod G_a) \bmod J]$ of all cluster center memory modules will be used.

5.3.4 Label Buffer

A FIFO label buffer is used to store the labels of N patterns as illustrated in Fig.5.10. When a new label is generated and pushed into the buffer, the corresponding old label will be popped out. These two labels are then compared. Since the label is represented as an integer, the comparator does not need to be pipelined. If the old label is different from the new label, a cluster membership change has occurred and the *label-change counter* will be incremented by one. At the beginning of each pass, the counter will be reset to zero. At the end of each pass, the value of the counter indicates the

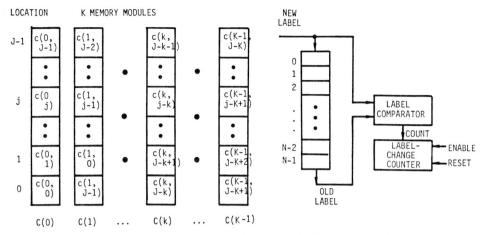

Fig.5.9. The skewed storage format of K cluster centers in K parallel memory modules

Fig.5.10. The organization of the pattern label buffer and the label-change measurement

total number of patterns that have changed their membership. This value will be used to determine whether the convergence has been reached. Usually, a predefined threshold is used. If the value of the label-change counter is greater than the threshold, another pass of evaluation will occur.

5.3.5 Cluster Counters

There are K counters to record the number of patterns assigned to K clusters in Fig.5.11. These counters are reset to zero at the beginning of each pass. When a new label is generated, the corresponding cluster counter is incremented by one. These values of the counters, $S(0)$ to $S(K-1)$, are used to calculate the coordinates of the new cluster centers as shown in Fig.5.5 and indicated in (5.2). Also, they can be used to identify outliers.

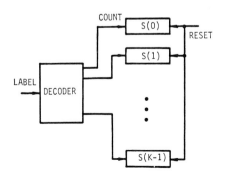

Fig.5.11. The organization of the cluster counters

5.4 System Operating Characteristics

5.4.1 Label Reassignment Process

For the label assignment process, NK Euclidean distances must be calculated between N patterns and K cluster centers according to (5.3). The evaluation of each distance is independent of the others except that they share the same data. Thus, there are NK vector reduction operations. The DSA cell depicted in Fig.5.6 is dedicated to the distance calculation. In order to utilize the DSA cells efficiently, each cell processes N vector input pairs. The DSA cell can generate G_a scalar values per $G_a J$ cycles using the interleaved multiple vector reduction technique as described in Sect.5.3.1.

For each pattern, we want to find the minimum distance among the K distances to the cluster centers. This requires N vector reductions for N patterns according to (5.6). A pipeline comparator is shared by the N vectors. The output of the comparator is the label (or the index of the cluster center) of the pattern, whereas the outputs of the DSA cells are the distances to each cluster center. A *label counter* is used to generate the corresponding index of the cluster center. The value of the index is appended to the distance generated by the DSA cells as shown in Fig.5.5. Thus, the input to the comparator is a distance and index pair. The distance is used in the comparison process and the index of the minimum distance is produced as the label of the pattern. First, we have to determine the input data stream to the comparator, which is the output of the DSA cells. Then, we may determine the input streams to the DSA cells.

Imagine that the output of the DSA cells is a distance matrix of N rows and K columns. To facilitate multiple vector reduction in the pipelined comparator, the matrix is partitioned into $\lceil N/G_a \rceil$ disjoint submatrices. The size of each submatrix is G_a by K. The input stream to the comparator follows a column-major order within each submatrix as shown in Fig.5.8b. Figure 5.12 demonstrates the timing diagram of the input data to the comparator. Since each DSA cell generates G_a elements per JG_a cycles and J is greater than or equal to K, there will be no output from the DSA cells during $G_a(J-K)$ cycles. Again, we can imagine the distance matrix as a virtual matrix of size N by J. The virtual distance is chosen to be a very large value so that it will not affect the selection of the minimum distance. Thus, the comparator will generate G_a labels every $G_a J$ cycles.

In order to generate the output stream from the DSA cells as shown in Fig.5.12, the sequence of input stream to the DSA cells must be carefully designed. The pipeline accumulator is the only processing unit in the DSA

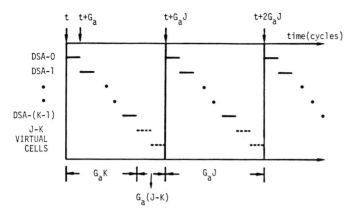

Fig.5.12. The timing diagram of the output distance data generated by the DSA cells, where $t = G_d + G_s + G_a J$

cell that has a feedback path. Since the accumulator has G_a segments, G_a vector input pairs can be processed at a time in an interleaved fashion. The input follows a column-major order as shown in Fig.5.8b. For interleaved multiple vector reduction, the DSA cell can generate G_a scalar values per $G_a J$ cycles. To generate a continuous output stream to feed the comparator, K DSA cells are needed. Each block of pattern matrix will flow through all K DSA cells. And each DSA cell is dedicated to calculate the distance to the corresponding cluster center. To avoid input conflict to the comparator, the input from the pattern matrix to each subsequent DSA cell must be delayed G_a cycles as shown in Fig.5.5. The delay can be easily implemented by using shift registers.

The reason for skewed storage of the cluster-center matrix should be clear now. Each element in the cluster-center matrix will be used repeatedly G_a times to operate with G_a patterns in a pattern submatrix. The skewing distance of 1 in the storage of the cluster-center matrix implies a delay of G_a cycles which matches the delay of G_a cycles of the pattern matrix input to the subsequent DSA cells.

Note that $J = \{\max M, K, G_a\}$. If J is less than K, the input to the comparator will result in a conflict because more than two DSA cells will simultaneously generate output data. Thus, J must be greater than or equal to K. Once the new label of a pattern is generated, the new label will be stored in the FIFO label buffer and will be compared with the corresponding old label as shown in Fig.5.10. The counter is incremented by one if the pattern has changed its membership. The value of the corresponding cluster counter is also incremented by one as depicted in Fig.5.11.

5.4.2 Cluster-Center Updating Process

The new cluster center is evaluated based on (5.2). Again, this is a vector reduction operation. However, the size of the vector may vary for different clusters. If all the patterns belong to one cluster, the size of that cluster is N. If a cluster has no patterns assigned to it, the size of that cluster is zero. The hardware implementation should be able to accommodate all possible data-dependent situations. For multiple vector reduction in a single pipeline, all vectors must be the same size. Furthermore, to facilitate pipelined processing, the vector input must be a continuous stream. Thus, each vector is considered to be size N, where each element is either a 0 or a real pattern data.

Recall that each pattern has M features. Evaluating a new cluster center is essentially performing a multiple vector reduction of M vectors, where the size of each vector is N. There are K cluster centers. Thus, we need K cluster-center evaluation units. According to (5.2), each unit consists of one pipeline accumulator and one pipeline divider as shown in Fig.5.5.

In the label reassignment process, the pattern matrix is treated as having N vectors with M elements each. However, in the cluster-center updating process, the pattern matrix is treated as having M vectors with N elements each. Thus, the pattern matrix must be fetched in row-major order. To allow overlapped processing of label reassignment and cluster-center updating processes, the pattern matrix must be treated as having J vectors with N elements each as shown in Fig.5.7. For interleaved multiple vector reduction of J vectors in a single pipelined accumulator, a dummy segment buffer is needed in the feedback path. The size of the dummy segment buffer is programmable and must be chosen to be $(J-G_a)$. Thus, the total number of segments in the accumulator is J.

Note that the pipeline comparator will generate G_a labels per $G_a J$ cycles. The buffer/decoder shown in Fig.5.5 has a FIFO buffer, which can store G_a labels, and a label decoder. One label is decoded at a time and will be held for J cycles. The decoded label will direct the pattern matrix input to the corresponding cluster-center evaluation unit. Other units will be fed with zeros. The selected unit will be fed with one row of J elements from the pattern matrix as indicated in Fig.5.8a. In order to synchronize the output of the pattern matrix and the output of the new label, a delay is needed with $G_s + G_d + G_a J$ cycles, to be explained later. After the sum of all the patterns in the kth cluster has been evaluated, the size of the cluster, S(k), is also known. The pipeline divider is needed to calculate the actual value of the new cluster center. If J is less than M, it will be very difficult to supply two input streams of the pattern matrix to facilitate overlapped processing between the label reassignment process and the cluster-center updating process.

5.4.3 System Performance Evaluation

In this section, we want to evaluate the total number of cycles required to perform one pass of label reassignment process and cluster-center updating process. In the first DSA cell, the external input will be fed into the pipeline accumulator after $G_d + G_s$ cycles. At the end of another $G_a J$ cycles, the first output will come out from the DSA cell and will be fed into the pipeline comparator. Another $G_a J$ cycles later, the first label will come out from the pipeline comparator. Let t indicate the processing time in terms of cycles. The first label will come out when $t = G_d + G_s + 2G_a J$.

The row-major access scheme of the pattern matrix is $G_a J$ cycles behind the column-major access scheme. Thus, a delay of $G_d + G_s + G_a J$ cycles is needed to synchronize the label input and the pattern input as indicated in Fig.5.5. Since there are $\lceil N/G_a \rceil$ blocks in the pattern matrix, $\lceil N/G_a \rceil G_a J$ cycles are needed to feed the first input to the pipeline divider. It takes another G_v cycles to perform the division and another M cycles to generate all M components of the new cluster center. The total processing time T(N,K,M) for one pass will be

$$T(N,K,M) = G_d + G_s + 2G_a J + \lceil N/G_a \rceil G_a J + G_v + M . \quad (5.9)$$

Usually, G_a is in the range of 3 to 5 and M and K are greater than G_a. If $J = M$, the total processing time is approximately $M(N + 2G_a + 1)$. Furthermore, if the number of patterns N is sufficiently large, the total processing time is approximately NM, which is the time required to fetch the whole pattern matrix once. If $J = K$ ($K > M$), the total processing time is approximately $K(N + 2G_a + M/K)$. Again, if the number of patterns N is sufficiently large, the total processing time is approximately NK, which is longer than the time required to fetch the whole pattern matrix once. As we mentioned in Sect.5.2, the value of M is in the range of 1 to 30 and the value of K is less than 20. Thus, in our design the total processing time of each pass is basically dominated by the time required to fetch the pattern matrix once in each pass.

Pattern clustering on a sequential processor is very slow. For an ideal sequential processor with no overhead on instruction scheduling and fetching, computing the distance between a pattern vector and a cluster center requires M subtraction operations (G_d cycles for each), M squaring operations

(G_s cycles for each), and M-1 addition operations (G_a cycles for each). To find the closest cluster center for each pattern, one needs (K-1) comparison operations (G_a cycles for each). To evaluate the new cluster centers, one needs M(N-K) addition operations and MK division operations. Therefore, the total number of cycles $T_s(N,M,K)$ needed for each pass on an ideal sequential processor is given by

$$T_s(N,M,K) = NK(MG_d + MG_s + (M-1)G_a) + N(K-1)G_a + M(N-K)G_a + MKG_v$$

$$= (NKM - N + MN - MK)G_a + NKMG_d + NKMG_s + MKG_v . \qquad (5.10)$$

If all the pipeline units have the same number of segments G, then (5.10) can be further simplified to (3NKM - N + MN)G. Under the same assumption and with J = M, (5.9) can be further simplified to 3G + 2GM + NM + M. When N is sufficiently large, the speed up of our design over the sequential processor is approximately 3KG. Note that this speedup is for one pass evaluation. If the instruction scheduling and instruction fetching overhead are also included in a sequential processor, a more significant saving in processing time has been achieved with our design.

5.5 Conclusion

We have presented a systolic architecture for squared-error pattern clustering. The systolic pattern clustering array can offload from the host the time-consuming pattern clustering tasks. The host will provide the initial cluster centers. Given a number of cluster centers, the systolic pattern clustering array will result in a partition of the patterns among clusters. Depending on the heuristic used, the host may add, delete, or merge the clusters and initialize another clustering task.

The processing time for one pass evaluation of the new cluster centers is dominated by the memory cycles required to fetch the pattern matrix once. It is a systolic architecture because data flows from the memory modules in a rhythmic fashion, passing through many processing units before it returns to the memory. It is a two-level pipelined design because each processing unit is also pipelined. The modularity and the regularity of the system architecture make it suited to VLSI implementations.

References

5.1 R.C. Dubes, A.K. Jain: "Clustering methodologies in exploratory data analysis", in *Advances in Computers*, Vol.19, ed. by M. Yovits (Academic, New York 1980) pp.113-228
5.2 S.Y. Lu, K.S. Fu: Stochastic error-correcting syntax analysis for recognition of noisy patterns, IEEE Trans. C-**26**, 1268-1276 (1977)
5.3 Q. He, R.C. Dubes: An experiment in Chinese speaker identification, in Proc. 1982 Int'l Conf. Chinese Language Computer Society (1982) pp.144-154
5.4 G.B. Coleman, H.C. Andrews: Image segmentation by clustering, in Proc. IEEE **67**, (IEEE, New York 1979) pp.773-785
5.5 J. Bryant: On the clustering of multidimensional pictorial data. Pattern Recognition **11**, 115-125 (1979)
5.6 G. Stockman, S. Kopstein, S. Benett: Matching images to models for registration and object detection via clustering. IEEE Trans. PAMI-**4**, 229-241 (1982)
5.7 R.K. Bleshfield, M.S. Aldenderfer, L.C. Morey: Cluster analysis software, in *Handbook of Statistics*, Vol.2, ed. by P.R. Krishnaiah, L.N. Karral (North Holland, Amsterdam 1982) pp.245-266
5.8 R.C. Dubes, A.K. Jain: Clustering techniques: The user's dilemma. Pattern Recognition **8**, 247-260 (1976)
5.9 K.S. Fu, T. Ichikawa (eds.): *Special Computer Architecture for Pattern Processing* (CRC Press, Boca Raton, Fla. 1982)
5.10 K. Hwang, K.S. Fu: Integrated computer architectures for image processing and database management. Computer **16**, 51-60 (1983)
5.11 H.T. Kung: Why systolic architectures? Computer **15**, 37-46 (1982)
5.12 K. Hwang, Y.H. Cheng: Partitioned matrix algorithms for VLSI arithmetic systems. IEEE Trans. C-**31**, 1215-1224 (1982)
5.13 H.T. Kung, C.E. Leiserson: Systolic arrays (for VLSI), in *Sparse Matrix Proceedings 1978*, ed. by I.S. Duff, G.W. Stewart (Soc. Indust. Appl. Math. 1979) pp.256-282
5.14 H.T. Kung, L.M. Ruane, D.W.L. Yen: "A two-level pipelined systolic array for convolutions", in *VLSI Systems and Computations*, ed. by H.T. Kung, B. Sproull, G. Steele (Computer Science Press, Roxville, MA 1981) pp.255-264
5.15 L.M. Ni, K. Hwang: Pipelined evaluation of first-order recurrence systems, in Proc 1983 Intern. Conf. on Parallel Processing, Bellaire, MI (1983) pp.537-544
5.16 M.R. Anderberg: *Cluster Analysis for Applications* (Academic, New York 1973)
5.17 K. Hwang, S.P. Su, L.M. Ni: Vector computer architecture and processing techniques, in *Advances in Computers*, Vol.20, ed. by M. Yovits (Academic, New York 1981) pp.115-197
5.18 L.M. Ni, K. Hwang: Vector reduction methods for arithmetic pipelines, in Proc. 6th Intern. Symp. on Computer Arithmetic, Aarhus, Denmark (1983) pp.144-150
5.19 P.P. Budnik, D.J. Kuck: The organization and use of parallel memories. IEEE Trans. C-**20**, 1566-1569 (1971)

6. VLSI Arrays for Syntactic Pattern Recognition[*]

Y.P. Chiang and K.S. Fu

During the past decades, the advanced IC technology has successfully reduced the size, cost and delay time of hardware devices. These reductions create or improve numerous areas of research and applications for computers. One of these areas is machine intelligence. In fact, machines with limited degrees of intelligence are already becoming an integral part of our lives. Examples range from sophisticated toys, health care, assembly automation and national defense. Pattern recognition techniques are among the most important tools used in the field of machine intelligence.

Recognizing a physical entity is one of the basic attributes of human beings and other living organisms. Yet, the process of recognition is very subtle and intricate. A pattern, as viewed by a computer, is a quantitative or structural description of an object of interest. The steps that a computer takes in order to recognize a pattern depend on the representation of that pattern and usually require a lot of computations. In this contribution, we briefly introduce pattern description methods, recognition procedures, and then concentrate on syntactic pattern recognition procedures and the influence of VLSI technology.

6.1 Pattern Description and Recognition

6.1.1 Pattern Description

There are many ways to describe a pattern. One typical approach is to convert each input pattern into a binary representation by means of a photosensitive matrix device. This measuring scheme is illustrated in Fig.6.1. Suppose we have a character pattern, shown in Fig.6.1a, which is projected onto a photocell matrix as indicated in Fig.6.1b, then the final binary pattern matrix will be obtained as the one in Fig.6.1c. The shaded cells are logical 1's

---
[*] This work was supported by the NSF Grant ECS 80-16580.

Fig.6.1. a) Character; b) measuring grid; c) resulting binary pattern

when they contain a sufficiently large character area and the white cells are logical 0's. The sensed data from the measurement grid is sometimes arranged in the form of a pattern vector

$$X = \begin{bmatrix} X_1 \\ X_2 \\ \vdots \\ X_n \end{bmatrix}.$$

In general, X_n is a binary number or a sampling value of a continuous function $X_n = f(t_n)$. It is from this pattern vector that one starts analysis and gathers important information about the pattern.

It has been observed that in some cases a pattern is a collection of many repetitive subpatterns in certain order. These subpatterns are called *pattern primitives*. The special order of primitive sequence is completely defined by the syntax of a grammar. For example, Fig.6.2a depicts the character "k". Figure 2b describes the pattern primitives. If we follow the definition of primitive interconnection operators [6.1] in Fig.6.3, the character in Fig. 6.2a can be described as ba + bb × aa × cc. In other words, patterns can be represented by strings of primitives. String representations are adequate for describing patterns whose structure is based on relatively simple connection of primitives. The more powerful approaches for many applications are realized through the use of high-dimensional representations such as trees, webs and graphs. However, in this chapter, we discuss only string representation.

6.1.2 Pattern Recognition

A pattern vector reflects the characteristics of a pattern in some degree. With suitable decision functions it is possible to distinguish pattern vectors into different classes. This type of pattern recognition procedure is usually referred to as the decision-theoretic approach [6.2]. This approach uses statistical decision and estimation techniques to classify patterns, but it exploits very little about the structure of the pattern. On the other hand, the string representation reveals all the structural information of the pattern. The recognition procedure of a pattern string is totally different

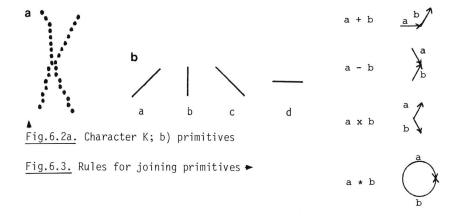

Fig.6.2a. Character K; b) primitives

Fig.6.3. Rules for joining primitives

from that of the decision-theoretic approach and is called the syntactic approach.

Every string is a primitive sequence. Similar patterns should have similar string representations. The collection of string is called *language* and it can be generated by a *grammar*. The grammar is a set of syntax rules which direct the formation of language. Every different pattern should be generated from different grammar. As an example, the character K from Fig.6.2a can be generated by the grammar G_k with production rules $S \rightarrow A + B \times C$, $B \rightarrow A \times D$, $A \rightarrow bA|B$, $C \rightarrow cC|c$, $D \rightarrow aD|a$. The other strings form this grammar like $b + bb \times a \times cc$ and $b + b \times a \times c$ are all legal characters of K. Therefore, syntactic pattern recognition becomes the problem of recognizing strings with a given grammar.

There are other problems involved in syntactic pattern recognition. For example, how do we infer a grammar from a set of sample patterns; how do we select and abstract primitives, etc? These problems are addressed elsewhere [6.3,4]. In this chapter we concentrate only on how to recognize a string with a given grammar.

6.2 Syntactic Pattern Recognition

In a syntactic pattern recognition system (SPRS) [6.4], an input pattern first enters the preprocessing stage, where it is encoded, smoothed and enhanced. Secondly, it enters the pattern representation stage and is converted into a string of primitives, such as context-free language (CFL). At the third and last stage, the input string is syntax analyzed, and the result shows whether the input pattern is rejected or accepted by the parser. In general the process of SPRS is very slow because of the tedious calculation involved in the parser. In the following we describe parallel CFL recognition algorithms which are suitable for VLSI implementation. This specially designed syntactic analyzer is fast and effective and therefore can speed up the whole process in SPRS.

The specific tasks involved in the syntactic analyzer are CFL recognition and error-correcting CFL recognition. The most commonly used recognizer in SPRS is *Earley*'s algorithm [6.5]. Although Earley's algorithm is suitable for general context-free language, it is a slow process with time complexity $O(n^3)$. Recently, *Graham* et al. [6.6] derived from Earley's algorithm a new on-line CFL recognition algorithm. This algorithm allows implementation only for $O(n^2/\log n)$ operations on bit vectors of length n, or $O(n^3/\log n)$ operations on a RAM. At the same time, *Weicker* came up with a similar result [6.7]: these two recognition algorithms are the fastest known. Earley's algorithm can be applied to error-correcting CFL recognition after we add the ability to count errors. Because of the repeated error checking and correcting, the error-correcting process has always been very slow. *Persoon* and *Fu* [6.8] proposed a sequential parsing scheme to reduce the computation load but not the complexity. To the authors' best knowledge, there is no parallel version of the error-correcting CFL recognizer. Before we discuss parallel recognition algorithms let us briefly review some basic definitions.

6.2.1 Definitions

Definition 1. A *context free grammar* is a 4-tuple $G = (V, \Sigma, P, S)$, where V is a finite set called the *vocabulary*, Σ is the set of *terminal* symbols or primitive symbols and $N = V - \Sigma$ is the set of *nonterminals*. The set of all finite length strings over V is V^*; P is a finite subset of $N \times V^*$ and is called the *rules* of G. The starting symbol is S.

Convention. Throughout this paper roman capitals A,B,..., denote elements of N, while lower case a,b,... are elements of Σ. But n will always be the length of the string and P is the set of production rules. Greek letters $\alpha, \beta, \ldots$ are elements of V^*, and λ denotes null string.

Definition 2. For any rule $A \rightarrow \alpha\beta$ we will call $A \rightarrow \alpha \cdot \beta$ a *dotted rule*. The dot "·" is a symbol not found in V and is used as a marker to indicate that the α part is consistent with the input string while the β part remains to be considered. Furthermore, $A \in N$ is called the *left-hand side* (lhs) of the rule, and $\alpha\beta \in V$ the *right-hand side* (rhs) of the rule.

Definition 3. If $X \in V$, the *predecessors* of $X = \{A | A \to X, A \in N\}$.

Definition 4. The X operator. Let Q be a set of dotted rules, then

$Q \times R = \{A \to \alpha U \beta \cdot \gamma | A \to \alpha \cdot U \beta \gamma \in Q, \beta \stackrel{*}{\to} \lambda$ and $U \in R\}$, if $R \subseteq V$.

$Q \times R = \{A \to \alpha U \beta \cdot \gamma | A \to \alpha \cdot U \beta \gamma \in Q, \beta \stackrel{*}{\to} \lambda$ and $U \to \delta \cdot \in R\}$, if R is a set of dotted rules.

Definition 5. The $*$ operator. Let Q and R be sets of dotted rules, then

$Q *R = \{A \to \alpha U \beta \cdot \gamma | A \to \alpha \cdot U \beta \gamma \in Q, \beta \stackrel{*}{\to} \lambda$ and there is some $U' \to \delta \cdot$ in R such that $U \stackrel{*}{\to} U'\}$.

Definition 6. Let $R \subseteq V$, define

PREDICT(R) = $\{C \to \gamma \cdot \delta | C \to \gamma \delta$ is in P, $\gamma \stackrel{*}{\to} \lambda$, $B \to C\eta$ for some B in R and some $\eta\}$.

In terms of dotted rule notation and the X and $*$ operators, Earley's algorithm is written as follows.

Algorithm 1

    t(0,0) = PREDICT({S});

    for j = 1 *to* n *do*

    begin [build col. j, given cols. 0, 1, ..., j-1]

    [Scanner:]

        for $0 \leq i \leq j - 1$ *do*

            t(i,j) = t(i,j-1) X $\{a_j\}$;

    [Completer:]

        for k = j-1 *down to* 0 *do*

        begin t(k,j) = t(k,k) $*$ t(k,j);

            for i = k-1 down to 0 *do*

                t(i,j) = t(i,j) U t(i,k) X t(k,j);

    end

    [Predictor:]

    t(j,j) = PREDICT($\bigcup_{0 \leq i \leq j-1}$ t(i,j));
    end .

Algorithm 1 constructs a recognition matrix $T = \{t(i,j)\}$. All the elements of T are sets of dotted rules. If $S \to \alpha \bullet \in t(0,n)$, then we can say that the input string has been correctly recognized.

Definition 7. A *c-dotted rule* is a dotted rule with error counter. It appears in the form $A \to \alpha \bullet \beta; k_A$, where k_A is the count of weights assigned to substitution (r), deletion (p) and insertion (q) errors. In the following definitions we assume $t(i,j)$ is a set of a c-dotted rules.

Definition 8. Suppose $A \to \alpha \bullet Ur; k_A \in t(i,k)$, $B \to \delta \bullet ; k_B \in t(k,j)$ and a_j is the jth input symbol, we define the "*'" operator as follows.

i) When $k = j-1$, $t(i,j-1)$ *'$\{a_j\}$ results in adding, if possible, $A \to \alpha U \bullet \gamma; k_A + X$ to $t(i,j)$, where $X = 0$ when $U = a_j$; $X = r$ when $U = b$ and $b \neq a_j$.

ii) When $i < k < j$ the $t(i,k)$ *' $t(k,j)$ is defined such that we add, if possible, $A \to \alpha U \bullet \gamma; k_A + k_B$ to $t(i,j)$ if $U = B$.

Definition 9. The operation $OPI(t(i,j))$ is defined as follows:

for all $A \to \alpha \bullet U\gamma; k_A \in t(i,j-1)$ and $t(i+1,j)$ add, if possible, $A \to \alpha \bullet U\gamma$; $k_A + q$ to $t(i,j)$.

Definition 10. The operation $OPD(t(i,j))$ is defined as follows:

for all $A \to \alpha \bullet U\gamma; k_A \in t(i,j)$, if $U \stackrel{*}{\to} a \in \Sigma$, then add, if possible $A \to \alpha U \bullet \gamma$; $k_A + p$ to $t(i,j)$.

Definitions 8-10 are formalized error-tracking procedures where r, p and q are weights assigned to substitution, deletion and insertion errors. With the help of these operations, it is easier to analyze the parallel algorithms in the following section.

6.2.2 Parallel CFL Recognition

Algorithm 1 is a sequential algorithm. Its computation order demands that no element of column $j+1$ can be processed until $t(j,j)$ is processed, and hence until all elements of column j are processed. This restriction is enforced by the predictor, which ensures that $A \to \alpha \bullet \beta$ appears in the ith row of the recognition matrix only if $S \to a_1...a_i A\gamma$. This restriction makes parallel execution of Algorithm 1 impossible. However, one can remove the predictor operation by replacing $t(j,j)$ with a constant set of dotted rules $Y = PREDICT(N)$ and still preserve the correctness of the recognition. The algorithm without the predictor operation is called the *weakened Earley's algorithm*. Notice that after replacing $t(j,j)$ with Y, there is no need to calculate the central elements $t(j,j)$ in Algorithm 1. Consequently, its $*$ operator has to be removed. In order that the algorithm may be correctly executed, we combine X and $*$ operators and define a new operator X^*.

Definition 11. The X^* operator. Let Q and R be sets of dotted rules, then

$Q \; X^* \; R = \{A \to \alpha U\beta\cdot\gamma | A \to \alpha\cdot U\beta\gamma \in Q, \; \beta \overset{*}{\to} \lambda \;\; \text{and} \;\; U \overset{*}{\to} \delta\cdot \in R\}$

and $\{B \to \delta C\xi\cdot\eta | \gamma = \lambda, B \to \delta\cdot C\xi\eta \in Y, \text{and} \; \xi \overset{*}{\to} \lambda, \; C \overset{*}{\to} A\}$;

if $R \subseteq V$, then $Q \; X^* \; R = \{A \to \alpha U\beta\cdot\gamma | A \to \alpha\cdot U\beta\gamma \in Q, \; B \overset{*}{\to} \lambda, \; U \in R\}$

and $\{B \to \delta C\xi\cdot\eta | \gamma = \lambda, \; B \to \delta\cdot C\xi\eta \in Y, \text{and} \; \xi \overset{*}{\to} \lambda, \; C \overset{*}{\to} A\}$.

With the X^* operator and the introduction of Y, Algorithm 1 will change its computation order to the one shown in Fig.6.4. Based on this unrestricted computation, we are able to write a parallel algorithm.

Algorithm 2

for $i = 1$ *to* n *do in parallel*

  $t(i-1, i) = Y \; X^*\{a_i\}$;

for $j = 2$ *to* n *do*

for $i = 0$ *to* n-j *do in parallel*

begin

[Scanner:]

  $t(i, i+j) = t(i,i+j-1) \; X^*\{a_{i+j}\}$;

[Completer:]

for $k = 1$ *to* j-1 *do in parallel*

  $t(i,i+j) = t(i, i+j) \cup t(i, i+k) \; X^* \; t(i+k, i+j)$;

end .

Algorithm 2 also constructs a recognition matrix $T = \{t(i,j)\}$. This time a dotted rule $A \to \alpha\cdot\beta \in t(i,j)$ if and only if $\alpha \to a_{i+1} \ldots a_j$. As proved in [6.9], this algorithm preserves the correctness of recognition. It is not difficult to see that Algorithm 2 takes time 2n to complete T if the input string has length n.

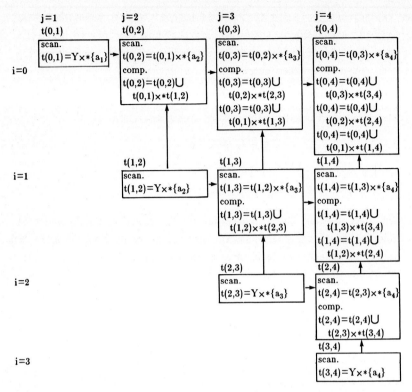

Fig.6.4. Computation flow of algorithm 2 (string length = 4)

6.2.3 Parallel Error-Correcting CFL Recognition

In this chapter we assume that only three types of errors can occur in a string. They are insertion, deletion and substitution. An error could result whenever a terminal symbol is encountered. We will not allow errors that involve a whole string of terminal symbols to occur. The string of terminals usually is derived from a nonterminal symbol.

In general there are two approaches in error-correcting recognition. The first approach [6.10,11] is to add production rules which generate all possible errors to the given grammar, and thus form a covering grammar. Then the erroneous input can be parsed with respect to the covering grammar in a normal way. Errors are counted whenever the error production rules are applied.

Another approach [6.12,13] is that of determining the minimum number of errors which could transform the input string into a valid sentence of the language being analyzed. In this chapter we adopt the second approach because of its straightforward procedure without introducing covering grammar.

Earley's algorithm can be applied to error-correcting recognition as long as proper error counting methods are included. Throughout this section, $t(i,j)$ will be a set of c-dotted rules. Definitions 8-10 had defined error tracking procedures for substitution, insertion and deletion errors. From Definition 11, operator X^* includes a constant set Y which has not been considered in Definitions 8-10. The error-tracking procedure for Y, in this case a constant set of c-dotted rules, is defined below.

Definition 12. The operation of $OPC(t(i,j))$ is defined as follows:

For all $A \to \alpha \cdot; k_A \in t(i,j)$, add if possible $\{B \to \delta C \cdot \eta; k_A | B \to \delta \cdot C\eta; 0 \in Y$, and $C \overset{*}{\to} A\}$.

Note that the statement "add, if possible", in Definitions 8-10, 12 indicate that the minimum error criterion was applied here. With these definitions we are ready to formulate the parallel error-correcting CFL recognition algorithm.

Algorithm 3

```
for i = 1 to n do in parallel
begin
    t(i-1, i) = Y *' {a_i};
    OPC(t(i-1,i));
    OPD(t(i-1,i));
end
for j = 2 to n do
    for i = 0 to n-j do in parallel
    begin
[Scanner:]
        t(i,i+j) = t(i,i+j-1) *' {a_{i+j}};
[Completer:]
        for k = 1 to j-1 do in parallel
            t(i,i+j) = t(i,i+j) U t(i,i+k) *' t(i+k,i+j);
        t(i,i+j) = OPI(t(i,i+j)) U OPC(t(i,i+j));
        OPD(t(i,i+j));
    end .
```

At the end of Algorithm 3, if $S \to \alpha\cdot;k_s$ appears in $t(0,n)$, we can say that the input string is recognized with k_s errors. Algorithm 3 constructs the recognition matrix $T=\{t(i,j)\}$, where $t(i,j)$ is a set of c-dotted rules, in the same manner as Algorithm 2 does. The only difference between these two algorithms is that the processes between every begin-end statement are different; in other words, Algorithm 3 also has time complexity 2n for input string of length n.

6.3 VLSI Implementation

Since the advent of VLSI technology, it is possible to put thousands of gates onto one chip. This reduces the cost of processors and increases the communication speed, but also changes the criteria for algorithm design. A good VLSI algorithm has to meet the following requirements [6.14,15]: (1) The function of each processor is kept simple and performs constant-time operation; (2) the communication geometry is simple and regular; and (3) the data movement is simple, regular and uniform. Above all, the algorithm should have a quite optimal use of the processing power present on silicon. Both Algorithms 2 and 3 have regular communication geometry as shown in Fig.6.4, and the direct VLSI implementation assigns each matrix element a processing cell. However, there are still two problems which need to be solved. The first one is that although every $t(i,j)$ within the same diagonal can be calculated simultaneously with the same time delay, the computation load in that diagonal would then be different from that of any other diagonal. The other problem is that basic operations in calculating $t(i,j)$ are data dependent, i.e., there is no constant execution time. These two problems are discussed further in the following sections.

6.3.1 Data Bus Arrangement

From Fig.6.4 we can see that in $t(0,2)$, the computation requires two data items, $t(0,1)$ and $t(1,2)$, plus one input symbol a_2; but in $t(0,4)$ six data items and one input symbol are needed. Therefore if Algorithms 2 and 3 are processed from diagonal to diagonal, there is no way to keep the same delay time for each individual diagonal. However, this problem was solved by *Guibas* et al. [6.16], whereby they proposed a VLSI array with an arrangement of slow and fast buses. This array was designed for optimal parenthesization which has a similar operation to that described in Fig.6.1. We added an input symbol bus (INP), which has the same traveling speed as a slow bus to Guibas'

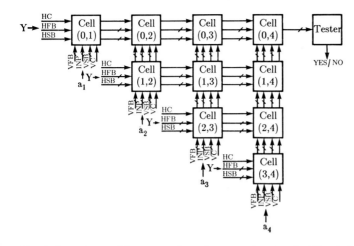

Fig.6.5. VLSI architecture for the computation in Fig.6.4

VLSI array; the resulting VLSI system, as shown in Fig.6.5, is capable of executing Algorithms 2 and 3.

In Fig.6.5 there are two control lines, horizontal control (HC) and vertical control (VC); two horizontal buses, fast bus (HFB) and slow bus (HSB); and three vertical buses, fast bus (VFB), slow bus (VSB) and input symbol bus (INP). The fast buses transfer data at a rate of one cell per unit time, while the slow buses and the INP bus have a rate of one cell per two unit times. Each cell has three functions, namely, loading the data onto fast buses, shifting the data from fast buses to the slow buses, and computing the "X" operator or "*'", OPD, OPI, OPC set of operations. The first two functions are essential for putting the right data in the right place at the right time, and they are controlled by VC and HC. The VC signal moves at a rate of two cells every three time units and controls the shifting operation from the fast buses to the slow buses. The HC signal moves one cell every two time units and is responsible for the fast bus loading operation. The timing sequence of this operation is shown in Fig.6.6 [6.16].

Next it is necessary to convert grammar rules into bit vector forms, and calculate Y, etc.; this information is then stored into every cell of the VLSI array. Starting from system time 1, all the cells and data buses are activated and the input symbols are read into the system in parallel, as shown in Fig.6.5; and the data buses in the central diagonal cells receiving 0's, except for HFB, are loaded with Y. This system runs in a pipelining and multiprocessing fashion. After 2n unit system times the tester receives data from cell (0,n) through HFB and after one more unit time, the tester will

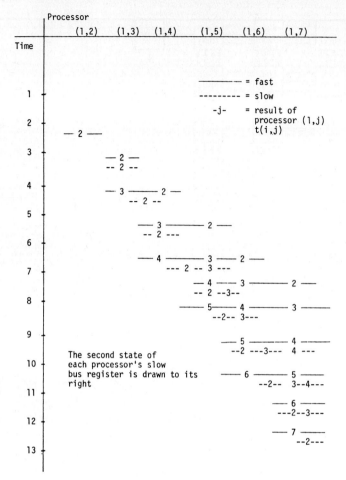

Fig.6.6. Time sequence for VLSI array

either tell whether or not the input string can be generated by the grammar (Algorithm 2) or give the error count (Algorithm 3).

The system operation described above is done under the assumption that the X* operator or the *' and OPC, OPD, OPI operations are data independent and can be executed in a constant time. To achieve this requirement, we need to reconsider the definitions of these operations.

6.3.2 Constant Execution Time

Since the operations in Algorithm 3 are the combination of operations in Algorithm 2 and the error-tracking procedures, we need to discuss only the constant execution time for X* operator. From Definition 11 we notice that λ pro-

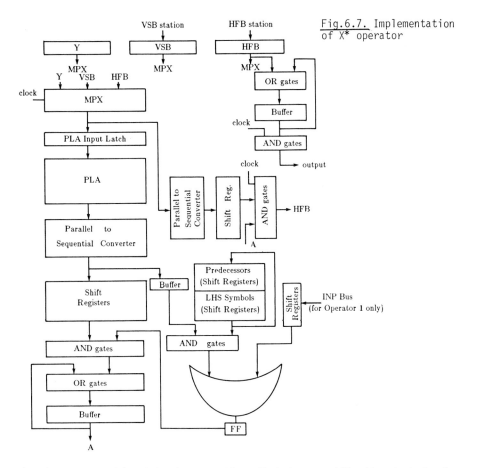

Fig.6.7. Implementation of X* operator

duction was considered in the operation; that is, we will add a dotted rule A → αUβ•γ instead of A → αU•βγ, provided $\beta \stackrel{*}{\Rightarrow} \lambda$. Since the length of β depends on the production rule, and the choosing of production rule depends on the data, this X* operator is data dependent. However, if we restrict our grammar to be λ free, the X* operator becomes data independent. Furthermore, with the help of bit-vector data structure, the X* operator can be implemented on the dedicated hardware, as illustrated in Fig.6.7, and executed in a constant time.

6.3.3 VLSI Array

The structure of a VLSI array which can execute both Algorithms 2 and 3 is shown in Fig.6.5. The differences are: (1) the data item which is transferred from cell to cell is different: one is a set of dotted rules and the other is a set of c-dotted rules; (2) the operations of each cell in the array are

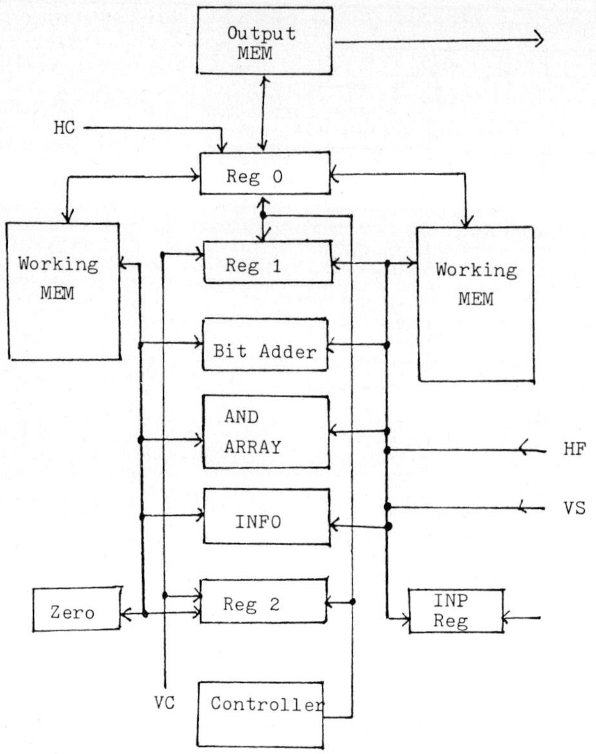

Fig.6.8. Cell structure for Algorithm 3

different; that is, besides loading data onto buses, the cell for Algorithm 2 computes X* while that for Algorithm 3 computes *', OPC, OPI and OPD.

The cell for Algorithm 2 can easily be assembled using two pieces of the special hardware indicated in Fig.6.7 plus one accumulator, five data bus registers and two control signal registers. Nevertheless, the cell operation for Algorithm 3 is too complicated to be economically implemented on any dedicated hardware. Therefore we chose the design in Fig.6.8 as the cell structure for Algorithm 3. This cell structure includes three general-purpose registers, sufficient working memory, a data bus register and limited special hardware. The detailed implementation is described in [6.9].

In these VLSI arrays, it is assumed that within one system unit time, every cell can complete its computation and every data bus can finish its data transfer operation. Although both Algorithm 2 and Algorithm 3 have time complexity 2n, the system unit time for Algorithm 3 should be longer than that of Algorithm 2.

These VLSI arrays require input data to be in parallel. In the case of Algorithm 2, the VLSI array generates a single output which indicates yes or no. The array for Algorithm 3 will generate the error counts besides the single bit indicator. The testers in both arrays have simple structure which contains registers and AND gates. There should be a *mask* stored in each tester which is used to determine whether the string is acceptable or not. This mask is expressed in a bit-vector form and it simply contains the grammar rules with S as LHS. In the following section we will explain the mask with a example.

6.4 Simulation and Examples

Simulation programs have been written and tested. These programs simulate the logical circuits in Figs.6.7,8, the cell, the VLSI arrays and the execution of Algorithm 2 and Algorithm 3. Several data have been tested and the results, as demonstrated in the following examples, demonstrate that our design is correct.

Example 1

The grammar [6.4] which generates strings describing median chromosomes is

$$G_m = (V, \Sigma, P, S)$$

where

$V = \{S,A,B,D,H,J,E,F,a,b,c,d\}$

$\Sigma = \{a,b,c,d\}$

and P:

$S \to AA \quad D \to FDE \quad H \to a$

$A \to cB \quad D \to d \quad J \to a$

$B \to FBE \quad F \to b$

$B \to HDJ \quad E \to b \quad .$

This grammar is an unambiguous context-free grammar. Since V has 12 elements, every symbol is represented by a 13-bit bit vector. There are 10 grammar rules and each rule has at most 3 RHS symbols, hence the cell data which represents $t(i,j)$ consists of 10 4-bit vectors. The grammar is coded into 10 groups of array of bit vectors. Each group has 4 13-bit vectors. Both the LHS symbols and their corresponding predecessors are composed of 10 13-bit vectors. Furthermore, $Y = PREDICT(N)$ has a cell data structure and its hexadecimal value is 8888888888. The mask residing in the tester has the value 2000000000 in hexadecimal which is a bit vector of the same size as a cell data. This mask indicates that S appears only in the first grammar rule as a LHS symbol.

After this information had been calculated and stored in the system, we tested several strings using Algorithm 2:

cbadabcbadab	YES	at time 25
cbadabccadab	NO	at time 25
cadacada	YES	at time 21
cbaaab	NO	at time 19 .

All the strings in this example have been correctly recognized. The required system time depends on the size of the system and the length of the string. In this example the system is a 12×12 triangular shaped array. The first two strings, with length 12, can be recognized at time 25, which is $2n+1$. Despite the fact that we have different recognition results, the two strings require the same system time.

The third string has length 8; therefore its correct recognition result is ready at time 16 on cell (0,8). However, this data has to travel four more system times in order to reach cell (0,12). Then, with one more system time, the tester can tell the recognition result from the data in cell (0,12). This is why the third string requires 21 system times. The system times required in this case follow the equation $2k+(n-k)+1$, where $k=8$ and $n=12$.

Example 2

The grammar [6.17] which generates the language $L(G) = \{a^i b^j c^k | i,j,k = 0,1,2,$ ... and $i = j$ or $j = k\}$ is given as

S' → e	S → D	C → cC
S' → S	S → C	C → c
S → AB	A → aA	D → aDb
S → DC	A → a	D → ab
S → A	B → bBc	
S → B	B → bc .	

Here we use e to indicate the null string (length = 0) which is included in the language. For convenience, we treat e as a special symbol which has length 1. Clearly, this grammar is ambiguous when generating strings of the type $a^i b^i c^i$. After preprocessing, several strings have been tested using Algorithm 2:

aaabbbccc	YES	at time 19
aabcc	NO	at time 15

100

aaa   YES   at time 13
e   YES   at time 11 .

In this example, the system has size 9×9; that is, it can recognize any string which has length no larger than 9. Again, all the strings have been correctly recognized.

Example 3

The median and submedian chromosome generating grammars [6.18] are used in this example. Six different strings which describe different chromosomes are tested, using Algorithm 3 under each grammar. The results are:

		median chro. error	submedian chro. error
1	cbabbdbbabcbabbdbbab	0	5
2	cbabbdbbbabbcbbabbbdbbaab	4	0
3	cadbbbabbcbbabbbda	6	6
4	ebbbabbcbbabbb	5	5
5	cbabbbbbabcbabbbbbab	2	5
6	cbabbbbabcbabbbbab	2	5

Each grammar in this example recognized its own string with error count 0. It also counts errors for the strings which do not belong to this grammar.

Example 4

Ten characters from Fig.6.9 are used in this example. Five of them are X's and the other five are K's. They are coded, according to PDL [6.1], into ten strings as listed below:

X's:
X_1: aa + cc × a × c
X_2: aa + cb × aa × cc
X_3: aaa + c × aa × b
X_4: aa + cc × aa × cc
X_5: aa + c × aa × cc

K's:
K_1: ba + bb × aa × cc
K_2: bb + bb × aa × cc
K_3: bb + bbb × aa × bcc
K_4: b + bbb × aa × bc
K_5: baa + bb × aa × cc .

The grammars which describe characters K and X have the following production rules:

G_K:
S → A + B × C
B → A × D
A → bA|b

G_X:
S → A + B × C
B → C × A
C → cC|c

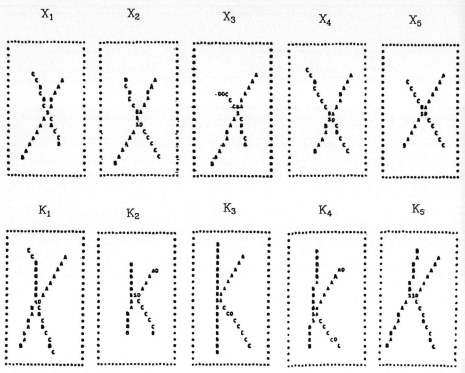

Fig.6.9. Ten characters

$C \rightarrow cC|c$ $A \rightarrow aA|a$
$D \rightarrow aD|a$.

The ten strings have been tested on both grammars. The resulting error counts are following:

	X_1	X_2	X_3	X_4	X_5	K_1	K_2	K_3	K_4	K_5
G_K	4	3	5	4	3	1	0	1	1	2
G_X	0	1	1	0	0	3	4	6	5	3

From the result it is clearly shown that Algorithm 6 can be used to classify patterns into groups of greater similarity.

6.5 Concluding Remarks

In this contribution we have presented parallel algorithms for CFL recognition, and error-correcting recognition. With the restriction that the input grammar has to be λ-free, all the three parallel algorithms have been simulated on triangular arrays, VLSI array and processor array, in a pipelining and multiprocessing fashion. All the array systems are assumed to be special functional modules of a host computer.

These arrays have their own system clocks. For an input string with length n, after $2n+1$ system time units, the recognition VLSI array distinguishes whether or not the input string is generated by the given grammar. If the input string is not accepted by the grammar, then after $2n$ times, $S \to \alpha \cdot$ will not appear in $t(0,n)$. The error-correcting recognition array provides the correct error count in $2n+1$ system time units. With a specified error threshold this error-correcting recognition array can be used to recognize noisy and distorted patterns.

All the arrays discussed require $n(n+1)/2$ cells for a string of length n. It is possible to reduce the hardware by using linear array with only n cells. But, this is not the best solution. In the linear array each cell is responsible for calculating one row in the triangular matrix. The parallel algorithms are still carried out diagonal by diagonal. Since the number of matrix elements decreases as we move along the diagonal, one cell in the linear array will be idle when we calculate the next diagonal. In other words, the first cell in the linear array will be busy all the time while the last cell functions only once. This is not very efficient. Besides, the linear-array approach increases the control complexity of each cell and makes the system expansion even more difficult. In the current design, the triangular array can process the next string after n unit times and the linear array has to wait until 2n unit times. Furthermore, if we use FIFO stacks as the bus registers, it is possible to process the next string every 2 units times in the triangular array, and the linear array has no room for pipelining processes at all.

The size of the array is critical to the performance. That is, an n by n upper triangular array can process strings only with length no longer than n. This is not a good parser for general language recognition. However, in syntactic pattern recognition applications, the string length of pattern can always be normalized to a fixed length or restricted within a certain range. Thus, the proposed techniques are very useful in syntactic pattern recognition. A universal array which can process strings with any length is now under investigation. This array uses two basic arrays, one triangular, the other square. Since a triangular array of any size can decompose into smaller triangular and square arrays, the main concerns in the universal array design are bus and control arrangements.

It is our belief that these parallel algorithms are applicable to many other fields such as compiler design [6.17,19] and multidimensional language parsing [6.20,21].

References

6.1   A.C. Shaw: A Formal Picture Description Scheme as a Basic for Picture Processing Systems. Info. and Contr. **14**, 9-52 (1969)
6.2   K. Fukunaga: *Introduction to Statistical Pattern Recognition* (Academic, New York 1972)
6.3   R.C. Gonzalez, M.G. Thomason: *Syntactic Pattern Recognition, An Introduction* (Addison-Wesley, Reading, MA 1978)
6.4   K.S. Fu: *Syntactic Methods in Pattern Recognition* (Academic, New York 1974)
6.5   J. Earley: An Efficient Context-Free Parsing Algorithm. Comm. ACM **13**, 94-102 (1970)
6.6   S.L. Graham, M.A. Harrison, W.L. Ruzzo: On Line Context-Free Language Recognition in less than Cubic Time, Proc. 8th Annual ACM Symp. on Theory of Computing (1976) pp.112-120
6.7   R. Weicker: General Context Free Language Recognition by a RAM with Uniform Cost Criterion in Time $n^2 \log n$, Tech. Rpt. No. 182, Computer Science Dept., Pennsylvania State University (1976)
6.8   E. Persoon, K.S. Fu: Sequential Classification of Strings Generated by SCFG's, Int'l J. Comput. Info. Sci. **4**, 205-217 (1975)
6.9   Y.P. Chiang: Parallel Processing and VLSI Architectures for Syntactic Pattern Recognition and Image Analysis, TR-EE 83-4, School of Electrical Engineering, Purdue University (1983)
6.10 K.S. Fu: "Error-Correcting Parsing for Syntactic Pattern Recognition", in *Data Structure, Computer Graphics and Pattern Recognition,* ed. by A. Klinger (Academic, New York 1976)
6.11 Q.Y. Shi, K.S. Fu: Efficient Error-Correcting Parsing for (Attributed and Stochastic) Tree Grammars. Inform. Sci. **26**, 159-188 (1982)
6.12 E. Tanaka, K.S. Fu: Error-Correcting Parsers for Formal Languages. IEEE Trans. C-**27**, 605-616 (1978)
6.13 G. Lyon: Syntax-Directed Least-Errors Analysis for Context-Free Languages: A Practical Approach. Comm. ACM **17**, 3-14 (1974)
6.14 H.T. Kung: Let's Design Algorithms for VLSI Systems, Proc. Caltech Conf. on VLSI (1979) pp.65-90
6.15 M.J. Foster, H.T. Kung: The Design of Special-Purpose VLSI Chips. Computer **13**, 26-40 (1980)
6.16 L.J. Guibas, H.T. Kung, C.D. Thompson: Direct VLSI Implementation of Combinatorial Algorithms, Proc. Caltech Conf. on VLSI (1979)
6.17 A.V. Aho, J.D. Ullman: *The Theory of Parsing, Translation and Compiling,* Vol. I (Prentice-Hall, Reading, MA 1972)
6.18 K.S. Fu: *Syntactic Pattern Recognition and Applications* (Prentice-Hall, Inglewood Cliffs, NJ 1982)
6.19 A.V. Aho, J.D. Ullman: *Principles of Compiler Design* (Addison-Wesley, Reading, MA 1977)
6.20 K.S. Fu, B.K. Bhargava: Tree Systems for Syntactic Pattern Recognition. IEEE Trans. C-**22**, 1087-1099 (1973)
6.21 W.C. Rounds: Context-Free Grammars on Trees, IEEE Conf. Rec. Symp. Switching and Automata Theory (1968)

Part III

VLSI Systems for Image Processing

7. Concurrent Systems for Image Analysis

G.R. Nudd

This contribution describes research on and development of real-time vision systems undertaken at the Hughes Research Laboratories over the past several years. In the course of this program our group has explored several areas germane to both software and hardware, our particular interest being the efficient and effective mapping of image analysis algorithms onto Very Large Scale Integration (VLSI) hardware. We describe here a number of processing techniques we are currently exploring including a number theoretic processor *Radius* [7.1], a systolic architecture TOPSS-28 [7.2], and a three-dimensional microelectronic processor [7.3]. In each case we have validated our concepts by operational hardware. The intent of this paper is to provide an overview of the problems and issues inherent in applying these VLSI technologies to image analysis and where necessary we refer to the existing publications for greater detail on each topic.

7.1 Background

Image analysis or computer vision has been a subject of interest for many years and it is well understood that a fully autonomous vision system would have major impact on a very wide range of applications. In the commercial world the automatic inspection of parts or the robotic assembly of complex machinery, for example, are areas of direct application and ones which are presently receiving a good deal of interest. More recently the military has become interested in exploring the potential of autonomous guidance and target identification applications based on computer analysis. To date two principal barriers have existed in the exploitation of these concepts: firstly the lack of a sufficiently robust set of algorithms for image analysis and secondly the need for significantly increased computational power.

The development of suitable software tools and techniques is a very substantial task. Even quite elementary operations such as line finding or the perception of texture prove to be quite complex for machine vision systems and can result in a relatively sophisticated code. This has led to a certain amount of unresolved discussion among the experts as to the most effective or elegant algorithm for many vision operations. As a by-product of this, most researchers in the field have preferred to write their software for general-purpose commercial machines where rapid changes can be effected and fine-tuning can be performed for each algorithm with a minimum cost. The penalty for this option of least commitment is usually computer execution time, and it is not atypical for a relatively small operation at the lowest end of the processing to require processing times ranging from many minutes to several hours on the best available machines. This is inconvenient and possibly costly, but for a researcher whose principal interest is in developing the optimum software it might be quite satisfactory.

However, these execution times are clearly unsuitable for many if not all the potential vision applications, and significantly more attention must be paid to efficiency and speed of processing. These issues which until now have not been of major importance will in effect determine whether vision will find widespread applications in the near future. Fortunately, we have recently seen substantial advances in both the software research for image analysis, and the processing capability of high-density microelectronic architectures. These two efforts have been pursued largely independently by experts in either software generation or processing technology. The advent of VLSI and efficient computer-aided design tools for system architecture development provides an opportunity to better integrate these two tasks. In this way algorithms and VLSI hardware can be developed to provide sufficiently efficient execution for real-time systems.

Two aspects seem to dominate the processing requirements: the need for greatly increased throughput, and the difficult issues associated with data and memory access. A study of the algorithms currently in use indicates that a computation rate in excess of 1,000 million operations per second will be necessary for real-time performance. This is at least an order of magnitude more than currently available with commercial machines.

One approach we have explored with some success is the use of special-purpose primitives for the low-level ionic tasks [7.4]. The concept requires developing a hard-wired VLSI component for each task or group of subtasks requiring high throughput (examples of this might be edge detection, line finding or local area convolution). These then work with a general-purpose host machine which performs the intermediate and high-level operations. An extension of this concept which reduces the range of VLSI primitives required is the development of a single programmable element which can perform the vast majority of low-level local area operations. An analysis of the high throughput operations indicates that many of them rely on a common set of basic operations such as local area convolution or sum of products. With this in mind we have developed a high-speed programmable local area operator which can be down-loaded from a host machine (currently a DEC computer). However, the complexity and computational rates required for such a processor are very high and we have explored a "number theoretic" approach to achieve them [7.5]. The resulting RADIUS machine [7.1] described below contains multiple copies of a custom NMOS chip to provide computation rates in excess of 200 million multiplications a second.

Another area receiving considerable interest recently is the application of 'systolic' architectures to achieve increased computational rates. These concepts are analogous to two-dimensional pipelined arrays in that the data flow is precisely mapped to avoid any processing delays incurred by memory fetch operations. As such they can be quite useful at the preprocessing level for operations such as enhancement or data compression. Like most pipelines they suffer from inflexibility as slight modifications are required in the computation, and our interest has been aimed at developing a general-purpose systolic element which, when configured in a variety of different arrays, can perform as wide a range of algorithms as possible. The resulting element or Multiplication Oriented Processor (MOP) is a VLSI chip with 15000 devices which is programmable and reconfigurable at both the chip and board level [7.2,6]. An initial application we are currently investigating is its use in spline calculations for bandwidth compression and efficient storage of synthetic and reference images [7.7].

The above examples use essentially conventional microelectronic techniques and their novelty derives either from the algorithmic mapping or the type of arithmetic used to implement the processing operations. In each case the principal emphasis is on exploiting parallelism and concurrency either at the algorithmic level (as in the systolic work, for example) or at the arithmetic level (as in the number theoretic process). However, we also review here another approach which goes beyond the present microelectronic technology and explores the use of three-dimensional microelectronic technology for image-analysis applications [7.3,8].

In Sect.7.2, we briefly review some of the processing constraints implicit in real-time image analysis and discuss a set of metrics for determining the effectiveness of various architectural configurations. Then in Sect.7.3 we describe the above architectural approaches, their performance characteristics and the VLSI hardware we have developed to date. Finally in Sect.7.4

we briefly review the outstanding issues in the development of autonomous vision systems.

7.2 Processing Requirements for Image Analysis

A concensus presently exists that some form of parallelism or concurrency will be required to provide the necessary computational power for image analysis, but as yet little has been done to quantify the capabilities of the various options suggested. One reason for the lack of any comparative analysis is the absence of a suitable set of metrics. In a recent paper [7.9] we suggested an analysis procedure which could provide a basis for evaluation and guide for future architecture work.

Conventionally two approaches are available for determining comparative performance. The first uses the instruction mix approach, where each program is examined to determine the number of times any particular arithmetic operation is used in the full procedure. Then the overall execution time is estimated from the product of this number and the execution rate for each instruction. For a given instruction mix this number is often taken as a performance criterion for a particular machine and various machines are compared on this basis. Unfortunately this approach is unsatisfactory for most image-analysis problems as it entirely ignores the issues of data access and management which are a dominant part of the processing cost. It is perhaps a better measure for serial processors where data has to be fetched from memory for each operation, but for the new parallel hardware where specific efforts are made to provide smooth and continual data flow at all processor sites it is insufficient.

Clearly an operation (such as local area filtering or convolution, for example) which requires data only from a local predetermined kernel requires much less memory overhead than an operation where the data may be widely dispersed or whose location depends on prior calculations (as in some relational processing). This point is significant as many of the recent advances in throughput for parallel VLSI architectures do little more than exploit a known data flow to provide a processing increase proportional to the number of processing elements (under optimum circumstances).

The benchmark approach avoids many of the problems and inaccuracies detailed above but requires careful choice of the representative programs. In general, the more sophisticated and elaborate the software becomes the more specialized and unrepresentative it is of the entire range of processing. A more satisfactory compromise is to determine inherent generic primi-

Table 7.1. Classifications of image-processing computations

- FUNCTIONAL STATISTICS
- LOCAL VS. GLOBAL
- MEMORY INTENSIVE VS. COMPUTATION INTENSIVE
- CONTEXT DEPENDENT VS. CONTEXT FREE
- ICONIC VS. SYMBOLIC
- OBJECT ORIENTED VS. COORDINATE ORIENTED

tives which are used at the low level and from which the high-level operations can be composed. Following the work of *Swain* et al. [7.10], who made a six-point classification scheme for general-purpose processing, we have identified an analogous set of criteria for image analysis as shown in Table 7.1. Most of the categories are common to general-purpose computation but the ionic versus symbolic distinction is particularly important and specialized to image processing. (Much of the high-level image analyis is currently being performed in symbolic form where data is stored and manipulated as lists rather than in direct image format.) The principal distinction occurs in the memory access times; in that for ionic processing there exists a direct physical relationship between the position within the image and memory location. This rather direct memory management facility does not exist at the more sophisticated level of the processing and data is accessed primarily by pointers to connected lists.

Table 7.2. Distribution of software metrics

OPERATION	LOCAL/ GLOBAL	LINEAR/ NONLINEAR	MEMORY INTENSIVE/ COMPUTATION INTENSIVE	OBJECT ORIENTED COORDINATE ORIENTED	CONTEXT FREE/ CONTEXT DEPENDENT	ICONIC/ SYMBOLIC
THRESHOLDING	L	NL	CI	CO	EITHER	I
CONVOLUTION	L	L	CI	CO	EITHER	I
SORTING	L	NL	MI	CO	CO	I
HISTOGRAMMING	G	L	CI	CO	CF	I
CORRELATION	L	L	CI	CO	CD	I
INTERIOR POINT SELECTION	G	NL	CI	OO	CD	I
LINE-FINDING	G	NL	EITHER	OO	CD	TRANSLATION
SHAPE DESCRIPTIONS	G	L	CI	OO	CD	TRANSLATION
GRAPH MATCHING	–	NL	MI	OO	CD	S
PREDICTIONS	–	NL	MI	OO	CD	S

Table 7.3. Classification of common architecture types

ARCHITECTURE	OPERATION CLASS												
	LOCAL	GLOBAL	LINEAR	NON-LINEAR	CONTEXT FREE	CONTEXT DEPENDENT	MEMORY INTENSIVE	COMP INTENSIVE	OBJECT ORIENTED	COORDINATE ORIENTED	ICONIC	SYMBOLIC	TRANSLATION
CELLULAR NUMERIC	++	+	++	0	+	0	-	+	--	+	+	--	--
PIPELINED	++	0	+	0	+	0	-	+	-	0	+	--	-
MIMD	++	0	0	0	+	+	+	+	0	0	+	0	0
NUMBER THEORETIC	++	++	++	-	++	-	0	++	--	+	+	--	--
SYSTOLIC	++	++	++	-	++	-	0	++	--	+	+	--	--
BROADCAST	+	-	0	0	0	0	+	0	0	0	0	++	0
DATA-DRIVEN	0	0	0	0	0	-	-	++	-	0	+	+	-
ASSOCIATIVE	0	0	0	0	0	0	0	-	-	0	+	++	0

++ HIGHLY SUITED TO APPLICATION
+ WELL SUITED
0 AVERAGE
- UNSUITED
-- HIGHLY UNSUITED

Hillis [7.11] analyzed the critical operations for symbolic computation and identified a set of principal operations including: deductions from semantic inheritance nets, matching operations for assertions, demons or productions, sorting according to chosen parameters and graph search and matching. In our classification these relational processing operations would be included as a part of graph matching.

A classification scheme such as that shown in Table 7.1 can be used to determine the optimum VLSI architecture for any given application from the underlying algorithmic primitive set. For example, once the generic operations are determined, the structure of the computational needs can be identified. In Table 7.2 we identify a commonly used set of software operations which might occur in scene analysis, for example, and their corresponding classifications. In Table 7.3 we use this classification to identify the relative strength and weaknesses of the common classes of architecture. As can be seen each architectural concept has advantages in some specific area but no single configuration is optimum for all systems. Therefore, the architectural choice might be made on the basis of the relative importance of the various primitives involved. The approach does emphasize the importance of clearly understanding the required processing operations prior to configuring the system and, more importantly, assigning relative weights or cost factors for each operation within the system. If this can be done for each application an optimum architectural concept can be formed from a combination of primitives as determined by the mapping given in Table 7.2. Each primitive can act as a silicon subroutine hosted to a general-purpose processor [7.4].

7.3 Concurrent VLSI Architectures

An approach which can reduce the complexity and cost of building a specialized high-speed custom processor for each primitive operation is to provide a single high throughput and programmable processor which can perform a wide variety of functions when down-loaded from a host. For example, a single processor can be built to perform all the local area arithmetic operations.

The throughput required of such a machine can be calculated as

$$\text{MIPS} = (n \times n) \times I \times O \times N^2 \times F \times 10^{-6} \quad , \tag{7.1}$$

where $(n \times n)$ is the local kernel size, I is the number of instructions per pixel in the kernel, N is the image frame size in pixels, O is the number of operations per algorithm, and F is the frame rate.

For an image equivalent to television this can result in throughputs of the order of 100 to 1000 mops (millions of operations per second) for even relatively simply operations such as edge detection over a small 5×5 kernel. Further, the instruction mix here is relatively high in multiplications which are time consuming and require large amounts of silicon area.

7.3.1 A Local Area Processor

A technique we have explored is the use of high-speed lookup tables to perform the arithmetic operations including additions, multiplications, etc. The advantage of this approach is that typically lookup structures are easy to design and can be made with both high speed and high yield if the tables are relatively small. The concept involves using high-speed random access memory (RAM) which can be programmed to provide the required arithmetic result of the form f(a,b) from two input arguments a and b. To enable this the contents of the table can be simply down-loaded with any appropriate function from the system host. In this way the system can be made programmable by simply reloading the memories. Unfortunately, if this is done directly, the size of the necessary memory is of the order of 2^W, where W is the word length.

For a typical system an internal arithmetic accuracy of 20 bits might be required. This implies use of tables in excess of one million words. To avoid this we have employed a residue arithmetic representation which essentially allows the system word size (W) to be broken into a number of smaller words $(w_1, w_2 \ldots \ldots w_n)$ such that

$$W = w_1 \times w_2 \times w_3 \times \ldots w_n \quad . \tag{7.2}$$

In this case the size of the tables is reduced from 2^W to 2^{w_i}, etc. Hence, if as in our case, we desire an overall accuracy equivalent to 20 bits this can be provided by four parallel and essentially independent channels each of 5 bits each employing a very small and potentially high-speed table of the order of 2^5 data elements.

The mathematical formulation of residue arithmetic can be found elsewhere [7.5] but the technique in essence involves dividing the incoming data by a series of prime or relatively prime numbers and operating only on the fractional or residue parts. In our case we use bases of 31,29,23,19 (each below 5 bits and hence each residue is constrained to be less than 5 bits). The resulting dynamic range is given by the product of the bases ($392863 = 2^{18.6}$). A schematic is shown in Fig.7.1. Within each channel, the data equivalent to a single base is processed and the arithmetic proceeds in the conventional

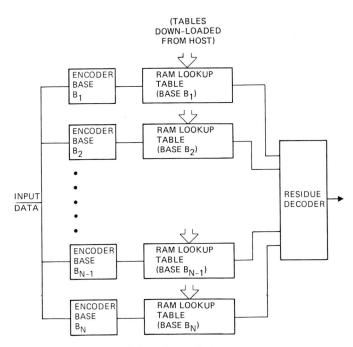

Fig.7.1. Residue arithmetic technique

manner with the proviso that as any number exceeds the base it is reconverted to the appropriate residue format by successive subtraction of the base. In this way no number within the channel exceeds five bits.

The final result in conventional number notation is produced by successively reducing the bases by a "mixed radix decoder" shown in Fig.7.2. A single custom NMOS chip consisting primarily of RAM cells with access times sufficient for 10 MHz operation was developed for the processor as shown in Fig.7.3. The system is built to perform arithmetic operations such as convolutions on a 5 × 5 kernel which involve 25 multiplications in each pixel interval. This translates to a total of 100 parallel lookups, one for each of the four bases, for each of the 25 pixels in the kernel. In this initial phase, to avoid the design complexity and inherent yield issues with VLSI, a total of twenty identical copies of a custom designed chip each with approximately 7,000 devices are used as shown in Fig.7.4. However, the full system could be designed in a single VLSI chip with approximately 80,000 devices.

The architecture is optimized for operations of the form

$$y = \sum_i f_i(x_i) \; , \tag{7.3}$$

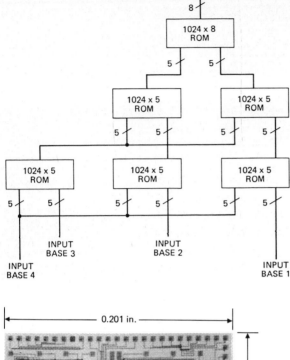

Fig.7.2. Mixed radix decoding circuit

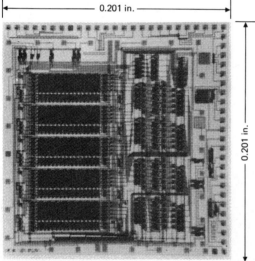

Fig.7.3. Custom chip for residue processor

where y is the output function, f is a polynomial function of a single variable and x_i are the pixels of the 5 by 5 element kernel. These are of particular importance at the low-level end of the processing and include all the linear two-dimensional filtering operators, etc. Because of the limitations of the residue number system not all arithmetic operations are permissible, and in particular division operations are excluded. However, this proves to

Fig.7.4. Full RADIUS processor

Table 7.4. Functional capabilities of RADIUS processor

POINT OPERATIONS
    POLYNOMIAL FUNCTIONS
    CONTRAST ENHANCEMENT

1— DIMENSIONAL OPERATIONS
    INTEGER COEFFICIENT TRANSFORMS
    POLYNOMIAL FUNCTIONS

2— DIMENSIONAL OPERATIONS
    EDGE ENHANCEMENT
    STATISTICAL DIFFERENCING
    LOW-PASS/HIGH-PASS FILTERING
    SHAPE MOMENT CALCULATIONS
    STATISTICAL MOMENT CALCULATIONS
    INTEGER COEFFICIENT TRANSFORMS
    TEXTURE ANALYSIS

be of little consequence in image analysis. A partial list of the classes of operations which the machine can perform at real time rates is given in Table 7.4, which includes the principal functions required at the feature extraction level of image analysis.

7.3.2 A Systolic Array for Processing

Another area of recent interest which we have been investigating is the application of systolic architectures to image analysis. The sytolic approach introduced by *H.T. Kung* [7.12] is similar to a two-dimensional pipeline in that very high efficiencies can be obtained for specific applications. Although one-dimensional string matchers and convolutional operations have been built this way most work in this area has been concerned with the solution of complex matrix arithmetic problems. To date each new application has essentially required a new architecture and processing element. Since many of the image analysis problems, particularly, at the preprocessing level including enhancement, restoration and bandwidth compression levels, can be conveniently expressed in terms of matrix manipulations of various kinds, we have been developing a general-purpose matrix array. Our interest has been twofold; firstly, the development of a single VLSI processing element that could be used over as wide a range of applications as possible, and equally important, the development of a single two-dimensional array architecture which would solve the full range of matrix applications.

We have therefore spent considerable effort in determining the basic computational operations required for a wide variety of array applications. The results of this work indicate that a single programmable bus-oriented processor with the functions indicated in Fig.7.5 will serve the vast majority of system applications. The key arithmetic element in the applications of this type is a high-speed multiplication and accumulate capability with some data manipulation facility as provided by the registers and LIFO stack. For this reason we have called our element a Multiplication Oriented Processor (MOP) [7.2]. Other important functions include square root and division as shown.

In our design we have used carry save arithmetic organized around an internal bus structure of 28 bits. The necessary bandwidth required for our applications is provided by two bidirectional, tristate, parallel ports with latches sufficient to receive or transmit two words per clock cycle. Two-dimensional arrays can be built by providing north, south, east and west communication. All the necessary control functions are embedded to allow maximum speed. The initial version of the processor using 4 micron design rules is shown in Fig.7.6. It contains approximately 15000 devices and completes a full multiplication and accumulate in approximately 1 μs using a 16 MHz internal clock. At present each chip contains only one processing element but potential exists, using one-micron design rules, to build a full 6×6 array on a single chip.

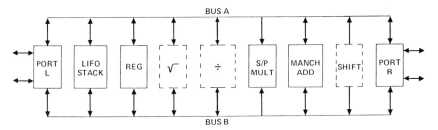

Fig.7.5. Principal functions required for general-purpose systolic element

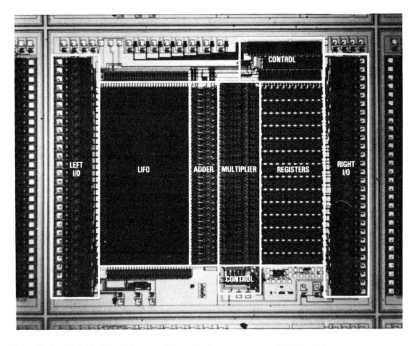

Fig.7.6. Multiplication orientated processor (MOP) chip

We have devised a general-purpose processor for two-dimensional data manipulation using an array of such elements as shown in Fig.7.7. The architecture is based on the *Faddeev* algorithms for in-place matrix solution [7.13], and can provide the range of processing options indicated by merely varying the organization of the input data. A key element of the concept as compared with the conventional systolic arrays is the elimination of the back-substitution array which usually has to be included and which typically disturbs the smooth data flow.

Initially we have developed a one-dimensional array of the MOP's element specifically for the solution of Toeplitz equations following the work of

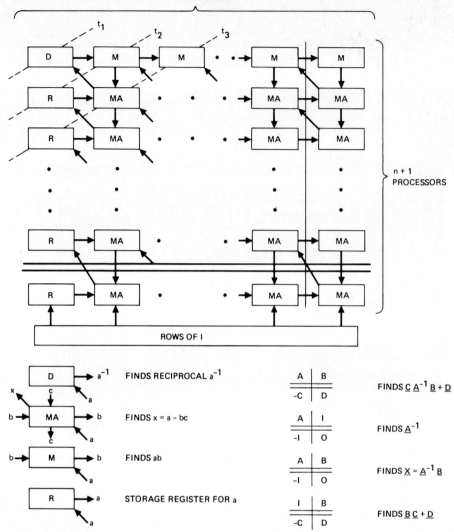

Fig.7.7. The Faddeev array for general matrix manipulations

Kung [7.14]. This requires only the linear architecture shown in Fig.7.8. The embedded algorithm follows the Weiner-Levinson solution of

$$R X = C \tag{7.4}$$

by triangular decomposition of **R** in an upper matrix **U** and its transpose U^+. In which case the solution is given by

$$X = U^{-1}D[U^+]^{-1}C \quad , \tag{7.5}$$

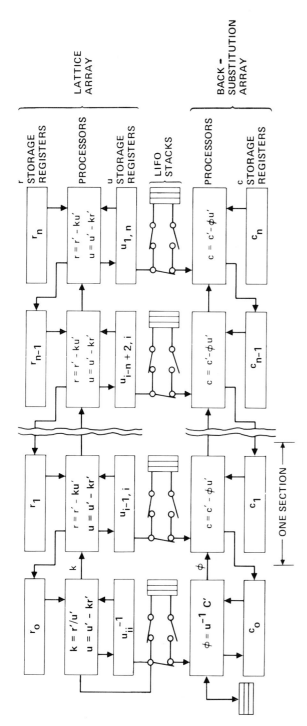

Fig. 7.8. Toeplitz System Solver TOPS-28

where D is the diagonal matrix containing the elements (U_{11}^{-1}, U_{22}^{-1}....). This involves only a series of Gaussian elimination steps by successive multiplication and accumulation and two back substitutions as indicated in the figure.

The Hughes Research Laboratories Toeplitz system solver TOPSS-28, although less comprehensive in its performance than the full Faddeev array, is nevertheless very useful in preprocessing applications for image analysis. For example, graphics and bandwidth compression techniques can be readily expressed in the form of a series of Toeplitz operations [7.7]. A case in point is the real-time calculation of splines for the smoothing of hand sketches or the reduction of data required for reference images. In a typical application a two-dimensional outline represented by a series of data points x,y can be stored in terms of a significantly reduced set of control points p_i, q_i. The original smooth data points x,y are related to the "compressed" data by the transformations

$$x(t) = \sum_{i=0}^{n} p_i N_{ir}(t) \tag{7.6a}$$

$$y(t) = \sum_{i=0}^{n} q_i N_{ir}(t) \tag{7.6b}$$

where N_{ir} are the precalculated spline functions. The data compression obtainable with this technique depends on the spatial bandwidth of the original image and of course the degree of fidelity required in the reconstructed data but it can range from less than ten to more than one hundred for smooth data. An example of five to one compression is given in Fig.7.9.

Equations (7.a,b) can be written in matrix form as

$$N[P,Q] = [x,y] \quad . \tag{7.7}$$

Since this is a rectangular system of matrix equations a unique solution for the control points in terms of the original data does not exist. However, the least-squares solution can be written

$$[P,Q] = [N^T N]^{-1} N^T [x,y] \triangleq N^- [x,y] \quad , \tag{7.8}$$

where N^- is the pseudo inverse.

For periodic spline functions, the matrix $N^T N$ above is either directly Toeplitz or nearly Toeplitz. For large graphics problems or complicated boundaries the solution of these equations in real time is not feasible in

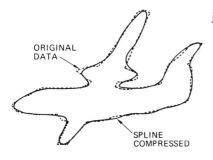

Fig.7.9. Example of image data compression using splines

part due to the difficulty of these inversions. The development of a one-dimensional systolic array such as TOPSS-28 or more significantly a general-purpose two-dimensional array as discussed above can provide a very effective approach to the problem and a valuable example of VLSI exploitation.

7.3.3 Three-Dimensional Architecture

Another architectural approach to image analysis which has received much interest over the past several years is the application of large cellular arrays. Here an array of identical processing elements is arranged, typically in a square tesselation with one processor for each picture element, and at each instruction cycle all the elements perform the same operation. The development of the Cellular Logic Image Processor (CLIP) [7.15], the Distributed Array Processor (DAP) and more recently the Massively Parallel Processor (MPP) [7.16] has demonstrated the power and advantage of this approach. These machines can perform with very high efficiency on two-dimensional data, particularly at the low and intermediate levels. For example, the recently fabricated MPP can provide a throughput of several billion fixed point instructions per second. The arithmetic in the machines considered to date is bit serial and hence potentially can have variable word length and provide some associative capabilities. The essentially slow serial processing at each site is more than compensated by the massive "word" parallelism.

The machines built to date have been major installations and are unsuitable for many real-time on-board applications. Recently, following the ideas of *Grinberg* et al. [7.3], interest has developed in applying high-density microelectronic technology to provide an ultra small high throughput cellular array. The cellular architectural configuration lends itself well to the new interest in wafer-scale technology and our work has shown that a very efficient machine can be mapped into a stack of microelectronic wafers. The three-dimensional microelectronic mapping provides very significantly advan-

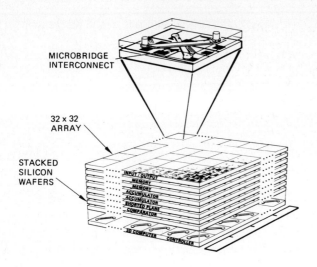

Fig.7.10. Three-dimensional computer

tages over the conventional two-dimensional layouts both in terms of the packing density and the average length of interconnect and therefore device size. Our systems level simulations over a wide range of applications, indicate a simultaneous improvement in power, volume and throughput of at least one order of magnitude. From our analysis this difference will enable a whole new range of both military and commercial publications to make use of image-based data.

The basic configuration of the machine is shown in Fig.7.10. It can be viewed as a stack of silicon wafers with a square $n \times n$ array of identical processors on each wafer. The architecture provides for only a relatively simple bit-serial circuit to be embedded at each cell site with perhaps 100 to 200 transistor complexity. In our present design using CCD and CMOS technology this results in a cell size of approximately 20 mils by 20 mils. These relatively simple cell designs allow high yield to be obtained with conventional processing. However, we employ a two to one redundancy at each cell site to achieve satisfactory yield across the entire wafer. In this case it can be shown, assuming uncorrelated silicon defects, an overall yield of

$$Y_T = P^M(2 - P^M)^{(n \times n)} \qquad (7.9)$$

can be obtained for an $n \times n$ array with M transistors per cell and a processing yield of P per transistor. From this expression it can be determined that given todays yield figures an effective 100% yield can be obtained with a block size of 250 devices on a 32 by 32 array.

The full processing capability of the machine is obtained by providing a separate wafer for each required logic function and assembling them vertically

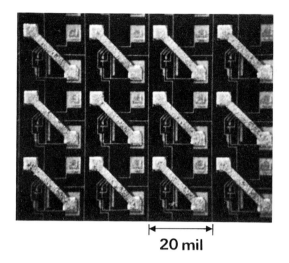

Fig.7.11. Processed wafer showing feedthrough and microbridge structure

Table 7.5. Component wafer types required in the 3D computer

CELL TYPE	FUNCTION
• MEMORY	• STORE, SHIFT, INVERT/NQN INVERT, "OR," FULL WORD/MSB ONLY, DESTRUCTIVE/NON DESTRUCTIVE READ OUT
• ACCUMULATOR	• STORE, ADD, FULL WORD/MSB ONLY, DESTRUCTIVE/NON DESTRUCTIVE READ-OUT
• REPLICATOR PLANE	• I/O, X/XY SHORT, STACK/CONTROL UNIT COMMUNICATION
• COUNTER	• COUNT IN/SHIFT OUT
• COMPARATOR	• STORE (REFERENCE, GREATER/EQUAL/LOWER
	• MULTI "AND" ON BUS LINE

so as to form an elemental microprocessor behind each pixel or data entry as indicated in Fig.7.11. We have found that a minimum of the seven wafer types listed in Table 7.5 are required for an efficient general-purpose machine. However, for specialized applications, it can be advantageous to include non-standard wafers.

Several underlying technologies have been developed to enable this configuration to be built efficiently and with high reliability. Principal among these is a means of providing communication at each cell site through the chip. For this we are currently using a proprietary process of thermal migration of aluminum to form an array (32 by 32 in our initial machine) of heavily p-doped vias or busses in n-type substrates. An equally important technology is the provision of high-quality mechanical and electrical contact between the adjacent wafers. Our approach here is to use a microbridge structure which makes electrical contact with the feedthroughs and spans across the active surface of the silicon as illustrated in Fig.7.11. Details of the technologies involved are given in [7.3]. We concentrate here on providing the basic concept of the three-dimensional processing operations.

The most obvious benefit from the cellular stack architecture occurs during the low-level or ionic processing. For these applications a full frame or subwindow of the image is fed into a single memory plane of the processor. Our concept employs a line feed which might typically be taken directly from a sensor or television camera and uses a serial-parallel approach to fill the uppermost plane. From this point, data is processed by moving the image up and down the stack to the appropriate processing plane using the parallel $n \times n$ array of buslines. The data flow and processing instruction stream is under the management of the control processor shown. This is typically a conventional von Neumann microprocessor through which instructions are entered. Each wafer contains an address decoder and instruction or configuration latch connected to a set of control lines at the edge of the array. Prior to executing any instruction each wafer ignores the system clock and does not communicate with the data busses. At the start of any operation the controller transmits the appropriate address, for each wafer involved, over the address bus. This has the effect of activating the required processors which then accept data from the control bus into an "on chip" instruction latch. This data carries the appropriate configuration code for the operations involved. In this way each wafer is configured sequentially prior to the start of the operation. The controller then commences the processing cycle by sending the appropriate number of clock cycles to the stack hardware. On completion of the processing the controller returns each wafer to the neutral state by activating the common reset line. The overhead in this process is about 18%.

A key element of the processor is the facility for data-dependent branching. One way this is achieved is by the use of a masking plane as illustrated in Fig.7.12. Essentially a template is formed corresponding either to a preassigned window or to a pattern resulting from preceding calculation. This binary pattern identifies only those data points which are required in a particular calculation and is used to activate a "wired and" circuit at each bus. The logical zeros have the effect of pulling the bus lines low and denying data access. Hence when a full frame of data is applied only those data points equivalent to the template are transmitted. This capability is

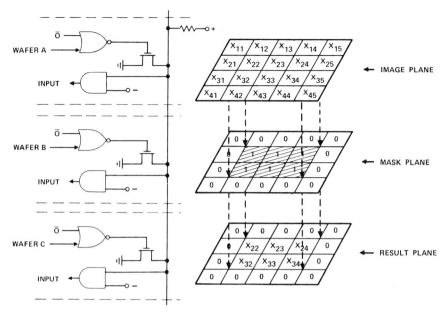

Fig.7.12. Use of masking plane to perform data-dependent branching

important to image analysis for such operations as local area extraction and interior versus exterior point selection, etc.

A circuit consideration which contributes to the solution of the potential power dissipation issues within the stack is the bit serial organization. Our present machine is configured for a sixteen-bit word length and a primary clock speed of 10 MHz. This results in an overall power dissipation measured in terms of the product of total number of gates and the clock speed approximately equivalent to the Department of Defense (DOD) Very High Speed Integrated Circuit (VHSIC) effort. Further, since the full energy in our machine is dissipated over a significantly larger area than a single chip, our thermal budgets are considerably more relaxed. Initial simulations indicate that a temperature variation across any wafer will be less than $5^\circ C$. However, to minimize the total power dissipation a combination of CCD and CMOS technology is used. For example, the memory cell shown in Fig.7.13a uses a 17-bit serial register for storage, so providing the basic storage capability in the processor. It has a communication path to its four most immediate neighbors and to the serial bus line for other wafers within the stack. It also provides capability for recirculation of the data, and for data inversion. The total device count for this configuration is the order of 150 transistor equivalents. A schematic of the accumulator cell is given in Fig.7.13b. It

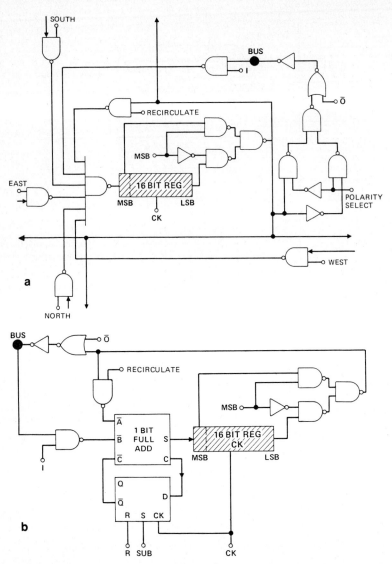

Fig.7.13a,b. Basic logic cells. a) Memory cell; b) accumulator cell

is similar to the above, but incorporates a one-bit serial adder as shown. In normal operation a 16-bit word is taken from the bus and summed a bit at a time to the data stored in the local register. Approximately 18 clock cycles or 1.8 μs are required for a full 16-bit addition.

The accumulator cell is also used for multiplication by performing the conventional shift and add algorithm requiring a total of 42 μs. A list of the basic arithmetic operations and their execution times is given in Table

Table 7.6. Execution times for basic arithmetic operations in the 3D computer

DATA MOVE	MEM	→ MEM	1.8 μS
ADD	ACC + MEM	→ ACC	1.8 μS
MULTIPLY	ACC × MEM	→ ACC	42.2 μS
DIVIDE	ACC ÷ MEM	→ ACC	127.1 μS
SQUARE ROOT	$\sqrt{ACC}$	→ ACC	152.6 μS

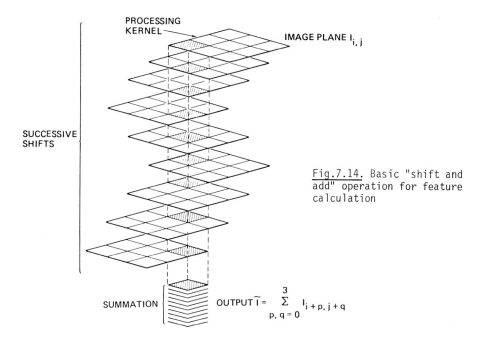

Fig.7.14. Basic "shift and add" operation for feature calculation

7.6. As can be seen, these speeds are not intrinsically fast for single operations but the large degree of parallelism within the array more than compensates for this.

In operation the HRL 3D processor performs much like the cellular architectures such as the MPP, DAP, or CLIP with principal exception of data access. A significant advantage of our approach is the essentially one-to-one access of the data to the individual processing cells. A simple example of this, which applies to all the two-dimensional filtering operations which form the basis of the edge detection and convolution functions, is illustrated in Fig.7.14. Here we illustrate the formation of a processing kernel (in this case performing a 3-by-3 local average) by performing a number of shifts and

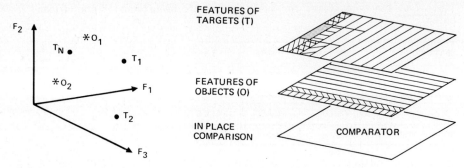

Fig.7.15. Parallel classification technique on the 3D computer

additions. This particular application involves only two planes; the original data and the summation plane. The full operation takes 100 μs and forms a local average for all picture elements simultaneously.

These "shift and add" operations, performed on a full array of data, can be extended to include all the two-dimensional linear filtering operations and many of the functions required at the feature extraction stage of image analysis. Nonlinear operations such as median filtering, formation of size filters and sorting operations can be performed by first calculating local histograms as described in [7.8]. Once these functions have been performed a typical classification operation can be undertaken by examining the least-square fit between the calculated features of the sensed objects and the prestored reference objects. The concept of the calculation is illustrated for a three-dimensional feature space by Fig.7.15. The object of the computation is to determine the least-vector distance between a set of reference vectors and the calculated features (typically the computation is performed on as many as 20 different features calculated). In principal the precalculated and the sensed features are repeated in row and columns respectively of two separate planes and an "in place" subtracted for all cell sites performed. A thresholding operation can then determine the least-vector difference representing the closest fit once a summation across all the features is performed.

The above operations are not intended to be in any way comprehensive but are given as examples of commonly used generic operations which are widely useful in image analysis. Details of a more comprehensive set of functions can be found in our previous publications and a more extensive publication [7.17] to appear.

7.4 Conclusions

In the foregoing discussion we have attempted to emphasize the need for parallelism and concurrency in the development of future VLSI architectures and illustrate the variety of options available to the system's architect. At present it is impossible to identify an approach which is optimum for all levels of image analysis and we have preferred to give examples which perform at each level of the computation. With this in mind our current thinking is that a composite processor may be preferred, consisting of a number of optimum, or near optimum processors, one for each level addressed in the calculation, all attached to a common general-purpose host and through which each is accessed. In this approach, the general-purpose machine would be responsible for the control and sequencing in addition to the more serial symbolic computations required at the high level. No doubt additional work will be undertaken on the application of VLSI to this problem, possibly resulting in a single machine optimized for the full range, but as yet no such development has appeared on the horizon.

References

7.1 S.D. Fouse, G.R. Nudd, A.D. Cumming: A VLSI Architecture for Pattern Recognition Using Residue Arithmetic, Proc. 6th Intern. Conf. on Pattern Recognition, Munich, FRG (1982)
7.2 J.G. Nash, G.R. Nudd, S. Hansen: Concurrent VLSI Architectures for Toeplitz Linear Systems Solution, Proc. Government Microcircuit Applications Conf., Orlando, FL (1982) pp.170-172
7.3 J. Grinberg, G.R. Nudd, R.D. Etchells: A Cellular VLSI Architecture for Image Analysis and Two-Dimensional Signal Processing, IEEE Computer Magazine
7.4 G.R. Nudd: Image Understanding Architectures, National Computer Conf. (Anaheim, CA 1980) AFIPS Conf. Proc. **49**, 377-390 (1980)
7.5 N. Szabo, R. Tanaka: *Residue Arithmetic and Its Application to Computer Technology* (McGraw-Hill, New York 1967)
7.6 S.S. Narayan, J.G. Nash, G.R. Nudd: "VLSI Processor Array for Adaptive Radar Applications," presented at SPIE 27th Intern. Techn. Symp., San Diego, CA (1983)
7.7 G.R. Nudd, J.G. Nash, S.S. Narayan, A.K. Jain: An Efficient VLSI Structure for Two Dimensional Data Processing, IEEE Intern. Conf. on Computer Design: VLSI in Computers, Port Chester, New York (1983)
7.8 G.R. Nudd: The Application of Three Dimensional Microelectronics to Image Analysis, in *Image Processing: From Computation to Integration*, ed. by S. Levialdi (Academic Press, London 1984) to be published
7.9 R.D. Etchells, G.R. Nudd: Software Metrics for Performance Analysis of Parallel Hardware, Proc. of Joint Darpa IEEE Conf. on Image Understanding, Washington, DC (1983)
7.10 P.H. Swain, H.J. Siegal, J. El-Achkar: Multiprocessor Implementation of Image Pattern Recognition: A General Approach, Proc. 5th Intern. Conf. on Pattern Recognition (Miami Beach, FL 1980) (IEEE Computer Soc. Press, New York 1982)

7.11 W.D. Hillis: The Connection Machine (Computer Architecture for the New Wave), MIT A.I. Memo No. 646 (1981)
7.12 H.T. Kung: Lets Design Algorithms for VLSI Systems, Proc. Conf. on VLSI Architectures, Design and Fabrication, California Institute of Technology (1979) pp.65-90
7.13 V.N. Faddeeva: *Computational Methods for Linear Algebra* (Dover, New York 1959)
7.14 S.Y. Kung: Impact of VLSI on Modern Signal Processing, Proc. Workshop on VLSI and Modern Signal Processing, University of Southern California (1982)
7.15 M.J.B. Duff: Review of the CLIP Image Processing System, Proc. of National Computer Conf. (1978) pp.1055-1060
7.16 K.E. Batcher: Design of a Massively Parallel Processor, IEEE Trans., C-**29**, 836-840 (1980)
7.17 G.R. Nudd, J. Grinberg, R.D. Etchells: The Application of Three-Dimensional Microelectronic Technology to Image Analysis. J. Parallel Distributed Computing (1984) to be published

8. VLSI Wavefront Arrays for Image Processing *

S.-Y. Kung and J.T. Johl

In image processing, there are tremendous demands for large-volume and high-speed computations. At the same time, the advent of VLSI offers computing hardware at extremely low cost. These factors combined are bound to have a major effect on the up-grading of future parallel image processors. As VLSI devices are getting close to their limits in performance, the solution to real-time digital image processing hinges upon novel designs of high-speed, highly parallel array processors for the common primitives in image processing, such as convolution, FFT, matrix operations, etc. This paper presents the algorithmic and architectural footing for the evolution of the design of these VLSI oriented array processors. With very large scale integration of systems in mind, a special emphasis is placed on massively parallel, highly modular, and locally interconnected array processors. Some image processing application examples of the array processor are discussed. One is the wavefront processing for least-square error solution with application to image reconstructions. The other is 2-D image correlation with application to image matching. As the trend of massively parallel system design points to the water-scale integration technology, we also discuss briefly some design considerations in implementing wafer-scale wavefront arrays.

8.1 Background

The ever-increasing demands for high-performance and real-time signal processing strongly indicate the need for tremendous computational capability, in terms of both volume and speed. The availability of low cost, high density, fast VLSI devices makes high speed, parallel processing of large volumes of data practical and cost effective [8.1]. This presages major technological breakthroughs in 1-D signal-processing and 2-D image-processing applications.

---

*Research supported in part by the Office of Naval Research under contract N00014-81-K-0191; and by the National Science Foundation under grant ECS-82-12479

It is noted that traditional computer architecture design considerations are no longer adequate for the design of highly concurrent VLSI computing processors. High VLSI circuit layout and design costs suggest the utilization of a repetitive modular structure. Furthermore, the communication has to be restricted to localized interconnections, as communication in VLSI systems is very expensive in terms of area, power and time consumption [8.1]. A broad and fundamental understanding of the new impact presented by the VLSI technology hinges upon a cross-disciplinary research encompassing the areas of algorithm analysis, parallel computer design and system applications. In this chapter, we therefore introduce an integrated research approach aimed at incorporating the vast VLSI computational capability into image processing applications.

8.1.1 Image-Processing Applications

A large number of image-processing algorithms are available today in software packages. For example, SPIDER, a transportable image-processing software package developed by the Electrotechnical Laboratory, MITI, and University of Tokyo, is a collection of 350 kinds of image-processing algorithms. However, the execution time of these algorithms, running on conventional computers, is often excessive and too slow for real application. In order to reduce the computation time drastically, it is necessary to resort to parallel processors, which takes advantage of the two-dimensional characteristics of the images and inherent concurrency of the algorithms. VLSI has made this parallelism economically feasible and technically realizable.

The problem really becomes: "How are these algorithms best implemented in hardware?" After examining most of the image-processing routines, some commonality becomes apparent—localized operations, intensive computation, matrix appearance. The architecture of an image-processing machine should reflect these considerations.

Let us examine some of the algorithms needed in a real-time, image-based visual inspection system. Its job is to recognize an object and check its dimensions against the specifications to see if the part should be accepted or rejected. The process involves algorithms in focus, location, tracking and measurement. A 2-D video sensor must first focus in on the scene, using an autofocus feedback system with an algorithm to determine when the image is the sharpest. Once the scene is in focus, the type of object must be determined. To accomplish this, the sensor image must be correlated with stored representations of objects in its database. Next, by analyzing the frame-to-frame position of the object, its velocity may be calculated to be used in tracking the object and keeping it within the field of view. Several algorithms can help improve the picture quality. If any optical distortion in the lens is known, this may be corrected with a geometric transformation. Also, if the sensors introduce any random noise, this can be reduced by image smoothing or median filtering. To determine the dimensions of the object, an outline is formed and then a straight line or curve is fitted to the edge. Finally, the outside measurements are checked against the specifications, and the part is either accepted or rejected.

In this real-time application with a rate of 20 frames per second and a 300×300 pixel image, over 10^6 pixels are processed every second. If each pixel requires 10 operations, a speed requirement of 10 MOPS is needed. It is quickly apparent that parallel processing help is needed.

8.1.2 VLSI's Impacts on Array Processor Architectures

The realization of many image-processing methods depends critically on the availability and feasibility of high-speed computing hardwares. In fact, until the mid-1960s, most image processing was performed with specialized analog (especially optical) processors because of the hardware complexity, power consumption, and lower speed of conventional digital systems. However, digital processors can provide better (and sometimes indispensable) precision, dynamic range, long-term memory and other flexibilities, such as programmability and expandability, to accommodate changing requirements. More importantly, today's VLSI offers a much greater hardware capacity, higher speed, and less power than, or at least compatible with, most existing analog devices.

Although VLSI provides the potential of tremendous increase in processing hardware, it imposes its own device constraints on the architectural design. Some of the key factors are communication, modularity, versatility, system clocking, etc. [8.2].

Interconnections in VLSI systems are often represented by two-dimensional circuit layout with few crossover layers. Therefore, communication — which costs the most in VLSI chips, in terms of area, time and energy — has to be restricted to localized interconnections [8.1]. In fact, highly concurrent systems require this locality property in order to reduce interdependence and ensuing waiting delays that result from excessive communication. Therefore, it is desirable to use pipelined array processors, with modular representations and localized interconnections, such as that in the systolic [8.3] or wavefront arrays [8.4].

Large design and layout costs suggest the utilization of a repetitive modular structure. In addition, it is very important to have flexibility in a processor module, so that the high design cost may be amortized over a broader market basis. There is, however, a trade-off between the flexibility of a processor and its hardware complexity. Our viable option seems to be that for a carefully selected applicational domain (image processing in this case), there may be significant commonality among the class of algorithms involved. This commonality can be identified and then exploited to simplify the hardware and yet retain most of the desired flexibility.

The timing framework is a very critical issue in designing the system, especially when one considers large-scale computational tasks. Two opposite timing schemes come to mind, namely, the synchronous and the asynchronous timing approaches. In the synchronous scheme, there is a global clock network which distributes the clocking signals over the entire chip. The global clock beats out the rhythm to which all the processing elements in the array execute their sequential tasks. All the PE's operate in unison, all performing the same, identical, operation. In contrast, the asynchronous scheme involves no global clock, and information transfer is by mutual convenience and agreement between each processing element and its immediate neighbors. Whenever the data is available, the transmitting PE informs the receiver of that fact, and the receiver accepts the data whenever it is convenient for it to do so. This scheme can be implemented by means of a simple handshaking protocol [8.2,4].

Since adopting an asynchronous, data-driven approach helps get around the necessity of centralized control and global synchronization, it is therefore desirable for the design of very large scale, highly concurrent computing structures. As we will show in Sect.8.5, it is especially appealing for reconfigured (or uncertain) interconnected networks such as wafer-scale integrated processors.

8.2 Parallel Computers for Image Processing

One of the main thrusts of the VLSI technology is the realization of supercomputer systems capable of processing thousands of MOPS or MFLOPS. Commercially available supercomputers now include the Cray 1/2, Cyber 205, and HEP from the United States, and Fujitsu VP200, Hitach S-810/20, and NEC SX-2 from Japan. However, from an image-processing point of view, the excessive supervisory overhead incurred in general-purpose supercomputers often severely hampers the processing rates. In order to achieve a throughput rate adequate for real-time image processing, the only effective alternative appears to be massively concurrent processing. Evidently, for special applications, (stand alone or peripheral) dedicated array processors will be more cost effective and perform much faster.

A reasonable compromise is to let supercomputers and (special-purpose) array processors play complementary roles in image-processing systems. In this case, the supercomputer will be the host computer supervising central control, resource scheduling, database management and, most importantly, providing super-speed I/O interfaces with the (peripheral) array processors. The concurrent array processors will, in turn, speedily execute the (compute-bound) functional primitives such as FFT, digital filtering, correlation, and matrix multiplication/inversion and handle other possible computation bottleneck problems.

8.2.1 FFT Array Processor

Since the invention of the FFT, a good majority of image-processing techniques uses the transform-based technique. This technique has drastically reduced computing time for many image-processing problems. The advent of VLSI devices will surely generate a new wave in the hardware implementation of (VLSI) digital FFT processors.

From another perspective, FFT based methods may not necessarily be the most attractive technique for certain image-processing problems, because of either the speed, convenience, or performance. The availability of low cost, high density, fast VLSI devices has opened a new avenue for implementing some other sophisticated algorithms.

8.2.2 Cellular Arrays

Advanced image-processing tasks, requiring excessive computational capability, often call for special-purpose designs. *Unger* [8.5] suggested a two-dimensional array of processing elements as a natural computer architecture for image processing. Such arrays are termed cellular arrays [8.6] when each of the PE's can directly communicate with its neighbors. A variety of image-processing algorithms have been developed for cellular arrays, such as local operations, histogramming, and moments and transforms.

As to the hardware construction of cellular arrays, the most advanced is perhaps the development of the Massively Parallel Processor (MPP) system by Goodyear Aerospace, to be used in processing satellite imagery at high rates. The MPP consists of 128×128 processing units, arranged in square array with nearest-neighbor interconnections. For 8-bit fixed-point image processing, the processing rates of the MPP can reach as high as 2000 to 6000 MOPS. However, for 32-bit floating-point computation, the processing rates are about 200 to 400 MFLOPS. In short, the MPP is most effective for applications that are computationally intensive and require relatively little input/output, such as processing satellite imagery or producing imagery from data sent by synthetic aperture radars.

8.2.3 Pipelined Array Processors

Apart from the algorithms such as FFT and sorting, we note that a great majority of image-processing algorithms possess inherent recursiveness and locality properties [8.4]. Indeed, a major portion of the computational needs for image-processing algorithms, except for FFT, can be reduced to a basic set of matrix operations and other related algorithms [8.7]. This fact should be exploited in the image processor architectural design. Obviously, the inherent parallelism of these algorithms calls for parallel computers with capacity much higher than any of the existing processor arrays. More challengingly, the inherent pipelinability of these algorithms should be incorporated in the design of new architectures. This in fact leads to the development of pipelined array processors [8.3,4,8].

The first major architecture breakthrough is represented by the two-dimensional systolic arrays for matrix multiplication, inversion, etc. Systolic processors are a new class of digital architectures that offers a new dimension of parallelism. The principle of systolic structure is an extension of "pipelining" into more than one dimension. According to *Kung* and *Leiserson* [8.9], "A systolic system is a network of processors which rhythmically compute and pass data through the system." For example, it is shown in [8.9] that some basic "inner product" PE's (Y <--- Y + A*B) can be locally connected to perform FIR filtering in a manner similar to the transversal filter. Furthermore, two-dimensional systolic arrays (of the inner product PE's) can be constructed to execute efficiently matrix multiplication, L-U decomposition, and other matrix operations, such as QR decomposition [8.10].

The basic principle of systolic design is that all the data, while being "pumped" regularly and rhythmically across the array, can be effectively used in all the PE's. The systolic array features the important properties of modularity, regularity, local interconnection, highly pipelined, and highly synchronized multiprocessing.

Another major development of pipelined array processors is along the line of wavefront arrays [8.4]. A wavefront array is a programmable array processor, which combines the features of the asynchronous, data-driven properties in data flow machines and the regularity, modularity, and local communication properties in systolic arrays. It has two distinct advantages: (1) it avoids the need for global synchronization; (2) it offers an effective space-time programming language.

While the original systolic concept focuses on the data movements between processors in the array, unfortunately, it does not naturally lend itself to a simple programming language structure. The need for a powerful description tool is further aggravated when more complex algorithms such as eigenvalue or singular value decompositions are encountered. Fortunately, the wavefront model utilizes the data-driven property of data flow machines and the locality and regularity of systolic arrays. This leads to a simple programming language for describing complex sequences of interactions and data movements.

8.3 Design of Pipelined Array Processors

To realize effectively the potential concurrency in a pipelined array processor, a new algorithmic design methodology will be needed. Concurrency is often achieved by decomposing a problem into independent subproblems or into pipelined subtasks. An effective algorithm design should start with a full understanding of the problem specification, the signal mathematical analysis, and the (parallel and optimal) algorithmic analysis; and then map these algorithms into suitable architectures. The effectiveness of (static) mapping of activities onto a processor array is directly related to the decomposibility of an algorithm. Moreover, the preference for regularity and locality will have a major impact in deriving parallel and pipelined algorithms. In our work, the two most critical issues—parallel computing algorithm and VLSI architectural constraint—are considered:

1) to structure the algorithm to achieve maximum concurrency and, therefore, the maximum throughput rate;
2) to cope with the communication constraint so as to compromise least in processing throughput rate.

To conform with the constraints imposed by VLSI, we shall now look into a special class of algorithms, i.e., recursive and locally (data) dependent algorithms. (In a recursive algorithm, all processors do nearly identical tasks and each processor repeats a fixed set of tasks on sequentially available data.) A recursive algorithm is said to be local type if the space index separations incurred in two successive recursions are within a given limit. Otherwise, if the recursion involves globally separated indices, the algorithm will be said to be global type; and it will always call for globally interconnected computing structures [8.11].

The assumption of recursive and locally data-dependent algorithms, however, incurs little loss of generality, as a great majority of signal-processing algorithms possesses these properties. One typical example is a class of matrix algorithms which are most useful for signal processing and applied mathematical problems. The concept of wavefront computing originates from algorithmic analysis; it will lead to a coordinated language and architecture design. In fact, the algorithmic analysis of, say, the matrix multiplication operations will lead first to a notion of two-dimensional computational wavefronts.

We shall illustrate the algorithmic analysis, leading to the wavefront concept and a coordinated architecture and language design by means of a matrix multiplication example. Let

$$A = [a_{ij}], \quad B = [b_{ij}],$$

and

$$C = A \times B,$$

all be $N \times N$ matrices. The matrix A can be decomposed into columns A_i and the matrix B into rows B_j and therefore,

$$C = [A_1 B_1 + A_2 B_2 + \ldots + A_N B_N］. \tag{8.1}$$

The matrix multiplication can then be carried out in N recursions,

$$c_{i,j}^{(k)} = c_{i,j}^{(k-1)} + a_i^{(k)} b_j^{(k)} \tag{8.2a}$$

$$a_i^{(k)} = a_{ik} \tag{8.2b}$$

$$b_j^{(k)} = b_{kj} \tag{8.2c}$$

for $k = 1, 2, \ldots, N$ and there will be N sets of wavefronts involved.

8.3.1 Pipelining of Computational Wavefronts

The computational wavefront for the first recursion in matrix multiplication will now be examined.

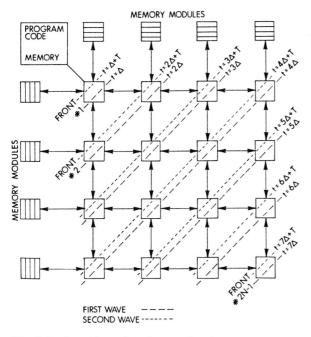

Fig.8.1. Two-dimensional wave-front array

A general configuration of computational wavefronts traveling down a processor array is exemplified in Fig.8.1.

The wavefronts are similar to electromagnetic wavefronts; each processor acts as a secondary source and is responsible for the propagation of the wavefront. Pipelining of the wavefronts is feasible because the wavefronts of two successive recursions will never intersect (Huygen's wavefront principle), as the processors executing the recursions at any given instant will be different, thus avoiding any contention problems. We note that the correctness of the sequencing of the tasks in the individual PEs is essential for the wavefront principle.

Suppose that the registers of all the processing elements (PEs) are initially set to zero:

$$c_{ij}^{(0)} = 0 \quad \text{for all } (i,j);$$

the entries of A are stored in the memory modules to the left (in columns), and those of B in the memory modules on the top (in rows). The process starts with PE (1,1):

$$c_{11}^{(1)} = c_{11}^{(0)} + a_{11} * b_{11}$$

is computed. The computational activity then propagates to the neighboring PE's (1,2) and (2,1), which will execute in parallel

$$c_{12}^{(1)} = c_{12}^{(0)} + a_{11}*b_{12} ,$$

and

$$c_{21}^{(1)} = c_{21}^{(0)} + a_{21}*b_{11} .$$

The next front of activity will be at PE's (3,1), (2,2) and (1,3), thus creating a computational wavefront traveling down the processor array. It may be noted that wave propagation implies localized data flow. Once the wavefront sweeps through all the cells, the first recursion is over (Fig. 8.1).

As the first wave propagates, we can execute an identical second recursion in parallel by pipelining a second wavefront immediately after the first one. For example, the (i,j) processor will execute

$$c_{ij}^{(2)} = c_{ij}^{(1)} + a_{i2}*b_{2j}$$

and so on. Here, let us be remined that it is possible to have wavefronts propagating in several different fashions. The only critical factor is that the order of task sequencing must be correctly followed.

8.3.2 Wavefront Architecture, Language, and Applications

With respect to architectural design, this data-driven feature is the key to get around the need of global synchronization—a potential barrier in the design of ultralarge scale systems. In the wavefront architecture, the information transfer is by mutual convenience between a PE and its immediate neighbors. Whenever the data is available, the transmitting PE informs the receiver of the fact, and the receiver accepts the data when it needs it. It then conveys to the sender the information that the data has been used. This data-driven computing scheme can be implemented by means of a simple handshaking protocol [8.4,12].

As mentioned earlier, the wavefront notion facilitates the description of parallel algorithms. The wavefront-oriented language, Matrix Data Flow Language (MDFL) as introduced in [8.4], is tailored towards the class of algorithms which exhibits the recursivity and locality mentioned earlier. The global MDFL [8.4] allows the programmer to address an entire front of processors. A preprocessor will then be used to translate the global MDFL language into (the local MDFL) instructions for the PE's. For a wavefront array the design of such a preprocessor is relatively easy since we do not have to consider the timing problems associated with the synchronous systolic array. A complete list of the MDFL instruction repertoire as well as some more complicated examples and the detailed syntax can be found in [8.4].

The notion of wavefront processing is applicable to all algorithms that possess recursivity and locality. It covers a broad range of the applicational algorithms, including most matrix operations (such as matrix multiplication, matrix inversion, and eigenvalue and singular value decomposition); signal- and image-processing algorithms (such as Toeplitz system solver, 1-D and 2-D convolution and correlation, recursive filtering and DFT); as well as some other scientific computations. An example of a (global) MDFL program for 2-D correlation will be given in the next section.

In summary, the top-down design methodology calls for closely coordinated design of algorithms, language and architecture. In wavefront array, the algorithm analysis dictates the language structure, which in turn determines the architecture of the computing network. The prospect of constructing high-speed (programmable) parallel signal processors, using VLSI wavefront arrays, appears to be very promising. In the following section, we shall discuss two examples of using wavefront arrays for image processing.

8.4 Image-Processing Applications

A special class of matrix algorithms, including those for singular value decomposition (SVD), least-square error solver, and 2-D convolution and correlation, have become popular in many image-processing applications. By successfully mapping the mathematical recursions of these algorithms into their corresponding computational wavefronts, these algorithms can be efficiently carried out by wavefront arrays.

8.4.1 A Least-Square Error Solver, with Applications to Image Reconstructions

In an image reconstruction problem, such as tomographical projection radiography, the cross-sectional imaging may be reconstructed by a stochastic estimation technique based on linear measurement, Fig.8.2. This problem may be mathematically formulated as

$$Y = AX + V \; ; \tag{8.3}$$

where Y = vector of measurements, X = object vector, A is the projection matrix, and V is the measurement noise vector.

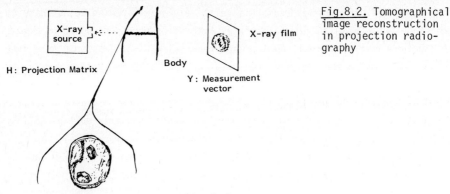

Fig.8.2. Tomographical image reconstruction in projection radiography

This amounts to solving an overdetermined system, and a popular approach is to use the least-square error solution,

$\hat{x}$: such that $\|Y - A\hat{x}\|_2$ is minimized,

where $\| \cdot \|_2$ denotes l_2 norm of the vector. Note that this often amounts to billions of operations. Therefore, fast and parallel methods have to be exploited for achieving any reasonable processing rates.

A numerically attractive method of solving the least-square error problem is the QR decomposition of matrix A [8.13]. (When the columns of A are linearly independent, then A can uniquely be decomposed into the product of two matrices, $A = QR$, where the columns of Q are orthonormal and R is an upper triangular matrix [8.13].) The least-square solution can now be obtained by applying:

$$Q^T A = \begin{bmatrix} R \\ 0 \end{bmatrix} , \quad Q^T Y = \begin{bmatrix} c \\ d \end{bmatrix} \qquad (8.4)$$

then solving:

$$Rx = c . \qquad (8.5)$$

This is the approach we will implement using the wavefront array. In doing so, we will make efficient use of a full array of processing elements.

The Givens Rotation

In this subsection, the Givens rotation will be utilized towards decomposing a matrix into its QR components to obtain the least-square error solution of a linear system. For all these, we need first to introduce a notion of computational wavefronts of Givens rotators. The main advantage of these operators is that they require only local communications for the generation of the rotation parameters. The Givens algorithm is based on applying an orthogonal operator $Q^{(q,p)}$, which performs a plane rotation of the matrix A in the (q,p) plane and annihilates the element a_{qp}. In all cases of applications to the wavefront array, the rotation is constructed based on adjacent elements of the processing array. When transforming a matrix into upper triangular form, the rotations are applied so as to annihilate the elements $a_{N,j}$, $a_{N-1,j}$, ... $a_{j+1,j}$, j = 1 ... N-1 in that order. Thus, the elements of the first column are dealt with first, then the elements of the second column, and so forth. For an upper triangularization procedure, we have:

QA = R ,

where R is an upper triangular matrix, and

$$Q = Q(N - 1)*Q(N - 2)* \ldots *Q(1) , \quad (8.6a)$$

and

$$Q(p) = Q^{(p+1,p)}*Q^{(p+2,p)}* \ldots *Q^{(N,p)} . \quad (8.6b)$$

$Q^{(q,p)}$ has the form:

$$Q^{(q,p)} = \begin{pmatrix} 1 & & & & & \\ & 1 & & & & \\ & & C(q,p) & S(q,p) & & \\ & & -S(q,p) & C(q,p) & & \\ & & & & 1 & \\ & & & & & \ddots \\ & & & & & & 1 \end{pmatrix} \begin{matrix} \\ \\ q-1 \\ q \\ \\ \end{matrix} \quad (8.7)$$

Columns: q-1, q. Rows: q-1, q.

such that

$$C(q,p) = \frac{a_{q-1,p}}{[a_{q-1,p}^2 + a_{qp}^2]^{\frac{1}{2}}}$$

$$S(q,p) = \frac{a_{qp}}{[a_{q-1,p}^2 + a_{qp}^2]^{\frac{1}{2}}} .$$

Note that $C(q,p) = \cos \theta$ and $S(q,p) = \sin \theta$, for some angle θ. The above operation of creating $\cos \theta$ and $\sin \theta$ is referred to as "Givens Generation" (GR). In fact, in a cordic implementation [8.2], this actually implies the computation of "cordic bits" for the rotation angle "θ" (Fig.8.3). The matrix product $A' = [Q^{(q,p)}]^{T*}A$ is then:

Array Size: Triangular Array (N x (N)/2)

Computation: $Q^T [A:Y] \longrightarrow [R:c]$

The least-square error solution of A X = Y will be

$$\hat{X} = R^{-1} c .$$

Initial: Matrix A is stored in the Memory Module (MM) on the top (stored column by column). Vector Y is stored in the last column.

Final: The result of R and c will be in the PE's.

```
WHILE WAVEFRONT IN ARRAY DO
BEGIN
   FETCH B, UP;
   CASE KIND =
   I. (GG) = ("GIVENS GENERATION")
REPEAT;
   BEGIN
      FETCH A, UP;
      GG(A, B,; B, θ );  (* MACRO INSR. *)
      FLOW θ, RIGHT;
   END;
   II. (GR) = ("GIVENS ROTATION")
REPEAT;
   BEGIN
      FETCH A, UP;
      FETCH θ, LEFT;
      GR (A, B, θ; B, C);
      FLOW C, DOWN
   END;
END;
```

Fig.8.3. MDFL programming for the triangularization in the least-square error solver

$$a'_{q-1,k} = C(q,p)*a_{q-1,k} + S(q,p)*a_{qk} \qquad (8.8a)$$

$$a'_{q,k} = -S(q,p)*a_{q-1,k} + C(q,p)*a_{qk} \qquad (8.8b)$$

$$a'_{jk} = a_{jk} \; ; \quad j \neq q-1, q$$

for all $k = 1 \ldots N$.

In the following, the operations in (8.8a,b) are referred to as Givens Rotation (GR). In the following, we shall discuss how to map the recursions into corresponding computational wavefronts in the array. Once $C(q,p)$ and $S(q,p)$ have been generated, they are propagated through rows q-1 and q of the array and influence those two rows only. Thus, the operation involves two distinct tasks: (i) GG: generation of the rotation parameters C and S, and (ii) GR: modification of the elements in the two impacted rows through the rotations of (8.8a,b). In general, the first task is carried out by the PE of the array that contains a_{qp} (the matrix element that is to be annihilated). The second task is executed by the remaining PE's in the same rows.

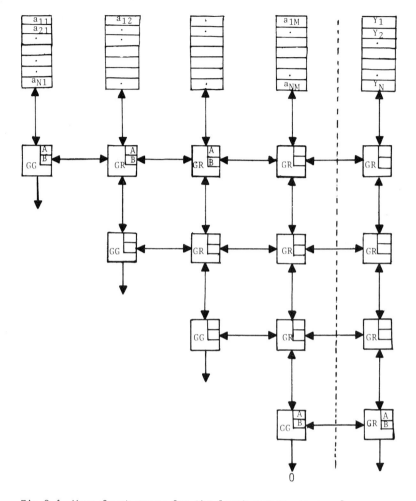

Fig.8.4. Wave-front array for the least-square error solver

The wave-front array takes a triangular configuration, as shown in Fig. 8.4, in which each PE includes a rotation processor.

The activities associated with operators $Q(1)$, $Q(2)$,..., etc., are described below.

Wavefront Processing for $Q(1)$ Operation

First, let us trace the fronts of activity relating to $Q^{N,1}$, i.e., the row operations involved in annihilation of the elements of the first column. The wavefront starts at PE(1,1), fetching a_{N2} from above and performing the computation for generating the rotation parameters $C(N,1)$ and $S(N,1)$ which annihilate a_{N1}. Upon completing this task, it will propagate the rotation parameters to PE(N,2), and then PE(N,3), etc., and further trigger each of them to perform the rotation operations as in (8.8a,b). (Note that one of the operands is fetched from above, and the updated result will be returned to the PE above.) Almost simultaneously, PE(1,1) is ready to fetch $a_{N-2,1}$ from above and [along with a'(N-1,1) created previously] generate the new rotation parameters corresponding to $Q^{N-1,1}$. Again, it will continue to trigger the successor PEs in a similar fashion. The processor for $Q^{N-2,1}$, $Q^{N-3,1}$, etc., also follow similarly.

In short, the computation activities are propagated sideways and they will further trigger the activity of PE(2,2), PE(2,3), etc., as discussed below.

Wavefront Processing for Q(2), Q(3),...,Operations

So far we have described the wavefront propagation only for Q(1). As soon as the elements a(N,2) and a(N-1,2) are updated by operations Q(N,1) and Q(N-1,1) in PE(1,2) and "flown" downward to the PE(2,2), the Q(2) wavefront may be triggered therein. The Q(2) wavefront, just like the Q(1) wavefront, will propagate steadily along the second PE rows. Similarly, the PE(2,3) will produce the two numbers and propagate them downward to PE(3,3) and initiate the wave front for Q(*,3), and so on. When all the N-1 wavefronts are initiated and sweep across the array, the triangularization task is completed.

Several important notes should be made here: (1) We shall first cascade A and Y, and apply Q^T to both of them, i.e., perform the operation $Q^T[A:Y]$, as shown in Fig.8.4. (2) Matrix Q is $N \times N$, R is a triangular $M \times M$ matrix, vectors c and x are $M \times 1$ and d is $(N-M) \times 1$. (3) In forming R from A, the Given Generator creates *cordic* bits for the rotation angle rather than C(q,p) and S(q,p), or $Q^{(q,p)}$. Applying the rotations on the row elements of A (resp. Y) is equivalent to executing A' = $[Q^{(q,p)}]^T*A$ (resp. Y = $[Q^{(q,p)}]^T*Y$). Thus, the final QR procedure outcome in the processor array will be the triangular matrix, R and c. (In doing so, Q need not be retained.) (4) The vector d represents the residual, and is output from the bottom PE of the rightmost column.

It is easy now to estimate the time required for processing the QR factorization. Note that for the least-square error solver, the second computational wavefront neet *not* wait until the first one has terminated its activity. In fact, the second wavefront can be initiated as soon as the matrix elements a_{N2} and $a_{N-1,2}$ are updated by the first wavefront. This occurs three task time intervals after the generation of that first wavefront.

(Here, for simplicity, it is assumed that generation of the rotation parameters as well as the rotation operation itself requires one task time interval.) Thereafter, the wavefronts for Q(1), Q(2),... can be pipelined with only three time intervals separating them from each other. The total processing time for the QR factorization would, therefore, be O(3N) on a triangular (i.e., half of N×N) array. An MDFL program for the triangularization of a matrix for the least-square error solver by the QR factorization is presented in Fig.8.3.

8.4.2 Wavefront Processing for Image Correlation

Image correlation is an important algorithm in the area of digital picture processing. This process is known also as picture matching, the problem of matching a pattern (e.g., a template) to a picture, or a piece of one picture to another [8.14]. A sequential computer may require minutes to handle the massive amount of computation needed for a moderately sized image. Some parallel-processing approaches have significantly helped in this area, but they suffer from high overhead and the need for lots of communication [8.15]. This necessitates a large interconnection network at an equally large expense.

An image is represented in the digital domain as a two-dimensional array of nonnegative numbers. Each number represents the intensity of a single pixel of the picture. Image correlation involves determining the position at which a small pattern f best matches a larger image g. This is done by shifting g into all possible positions relative to f; and, in effect, the template will move over the image as shown in Fig.8.5, and at each point the normalized cross correlation will be computed.

The computation involves

$$c(m,n) = \frac{\sum_i \sum_j f(i,j) \, g(i+m,j+n)}{(\sum_i \sum_j [g(i+m,j+n)]^2)^{\frac{1}{2}}} \quad \text{at each placement.} \quad (8.9)$$

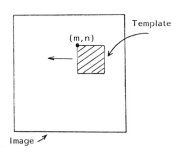

Fig.8.5. Moving window for image correlation

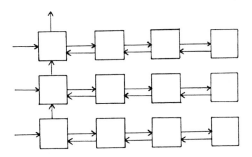

Fig.8.6. Wavefront array configuration for the correlation algorithm

The point at which this value is the largest determines where a template best matches the picture.

The wavefront array processor algorithm for image correlation will utilize the regular structure of the array. All operations will be pipelined in an attempt to keep all processors "busy." This algorithm assumes that if the match area is an $M \times N$ arrangement of pixels, then the WAP is configured as an $M \times N$ array of processors as shown in Fig.8.6, so that each processor can hold a single pixel of f, the template. The array will perform the integration and normalization needed for pattern matching. The image will pass through the array from left to right, and the correlation measure will emerge from the (1,0) processor. The template is stored in the processor array and appears stationary as the image data enters from the left and exits on the right in parallel motion.

In this subsection, we shall concentrate on the numerator in (8.9), i.e., the correlation function, and ignore the normalization factor [the denominator in (8.9)]. It can be shown that the normalization factor can be computed by following a similar procedure.

For simplicity of discussion, the recursive algorithm for one-dimensional image correlation (the i dimension is dropped) is first described

$$g_{j+1}^{(k)} = g_j^{(k)} \quad , \quad c_j^{(k)} = c_{j+1}^{(k-1)} + f(N + 1 - j)g_{j+1}^{(k)} \quad , \quad i = 1,\ldots,M \quad .$$

The correctness of this algorithm can be verified by taking the z transform over the wavefront index, i.e., the superscript (k). To simplify the proof, let us assume $N = 3$, then

$$G_{n+1}(z) = G_n(z)$$

$$C_n(z) = C_{n+1}(z) \cdot z^{-1} + f(4 - n)G_{n+1}(z) \quad . \tag{8.10}$$

Equation (8.10) implies that $G(z) \equiv G_n(z) \equiv G_{n+1}(z)$. Therefore,

$$\begin{aligned}
C_1(z) &= C_2(z) \cdot z^{-1} + f(3) \cdot G(z) \\
&= [C_3(z)z^{-1} + f(2)G(z)]z^{-1} + f(3)G(z) \\
&= [C_4(z)z^{-1} + f(1)G(z)]z^{-2} + [f(2)z^{-1} + f(3)]G(z) \\
&= [f_{(1)}z^{-2} + f_{(2)}z^{-1} + f_{(3)}]G(z) \\
&= F^*(z) \cdot G(z) \quad ,
\end{aligned}$$

which is the z transform of the correlation of f(n) and g(n).

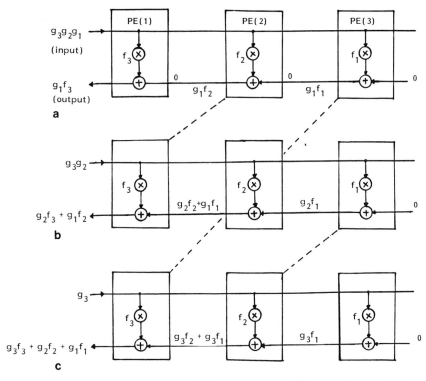

Fig.8.7. Illustration of the activities corresponding to the wavefronts for the correlation algorithm

To illustrate wavefront processing in one dimension, let the first front carry g_1 across the array. In Fig.8.7a, the first row (of a two-dimensional array processor) is shown. Since the initial values for the right operand of the adder are set to zero, $g_1 \cdot f_i$ will be output at all stages. Next, (as shown in Fig.8.7b) as the second front propagates, $g_2 \cdot f_i$ is produced; however, there are different interactions with the adders. For processor (1), $g_2 \cdot f_3$ will meet the input from the right (which is the output of the previous wave front), i.e., $g_1 \cdot f_2$, and result in an output $g_2 f_3 + g_1 f_2$. Similarly, the processor (2) will yield $g_2 f_2 + g_1 f_1$, ready to meet the third wave front in (1) processor. As the third front, carrying g_3 is started (Fig.8.7c), the processor (1) will yield $g_3 f_3 + [g_2 f_2 + g_1 f_1]$. The successive outputs will be the correlation function of the form $\sum_j g_{j+n} f_j$, with n (starting at one) indicating successive correlation measures.

In the two-dimensional case, the same procedure will be applied to the mth through $m + M - 1^{st}$ rows of the image simultaneously. Then all the M one-dimensional correlation functions (which are the output of each row) will be added together by the processors on the left-hand side. Finally, the complete sum will emerge from the top, giving

$$\sum_i \sum_j f(i,j) g(i + m, j + n) \quad .$$

After the template scans across the entire block of M rows, the m index is incremented by one and the same procedure is repeated.

The complete two-dimensional correlation algorithm for the WAP can be expressed in MDFL instructions as shown below. Assume that template data f has already been stored in register F. It will remain there throughout the algorithm.

```
BEGIN
    WHILE WAVEFRONT IN ARRAY DO
        BEGIN
            CASE KIND =
            (*, 0) = THE LEFT-MOST COLUMN PEs
                    BEGIN
                        FETCH A, LEFT
                        FLOW  A, RIGHT
                        FETCH C, DOWN
                        FETCH B, RIGHT
                        ADD B, C, C
                        FLOW  C ,UP
                    END;
            (*, *) NOT (*, 0)
                    BEGIN
                        FETCH A, LEFT
                        FLOW  A, RIGHT
                        MULT A, F, A
                        FETCH B, RIGHT
                        ADD A, B, B
                        FLOW  B, LEFT
                    END;
            ENDCASE;

        END;
END PROGRAM.
```

In summary, in handling image correlation algorithms, the wavefront array processor, an asynchronous and data-driven multiprocessor, would offer roughly the same computation time as an SIMD machine, but reduce the interconnection complexity. This regular array of processors is eligible for VLSI implementation. Furthermore, an integrated approach of language and architecture has helped make this multiprocessor system easy to use. With such a pipelined array, not only does the overhead decrease, but the program actually becomes simpler.

8.4.3 Other Applications

Image-processing operators (point-spread functions), unlike that in time series processing, are often noncausal operators. The deconvolution in the spatial domain will then involve solving Toeplitz systems with a special band-matrix structure. An efficient, fast and pipelined solution for such banded Toeplitz systems has been developed by *Kung* and *Hu* [8.16,17]. We also note that the applications mentioned above should not be restricted to time series or image processing only, they are equally useful for other multi-dimensional processing such as beam forming in azimuth and elevation, space-time wave-number processing, and adaptive processing of multichannel systems.

8.5 Wafer-Scale Integrated System

The massive processing hardware as required by many image-processing applications calls for novel implementation schemes. One very promising approach in the near future is the so-called wafer-scale integrated processor. With this technique applied to the wavefront array, many identical processors can be configured as a 2-D array on a single wafer [8.18]. Compared to conventional interconnection means, this technique offers large savings in speed, power, and cost.

Due to device processing defects there will inevitably be "faulty" processor elements on every wafer. Strategies must be devised to route around the faulty ones. These strategies involve minimizing the total interconnection path and maximizing the size of the good array.

Recent technological advancements have contributed to the feasibility of wafer-scale integration.

(i) E-beam technology enhances testability by making the internal circuitry more observable.
(ii) Laser programming makes and breaks interconnection paths between cells on wafer.

Since each wafer processed will have unpredictable defective cells, each reconfigured array will have its own unique interconnection pattern. This is bound to create more timing uncertainty, thus making the more flexible wavefront processing very appealing compared to the pure systolic processing. The example in Fig.8.8 illustrates the varying delay paths in a 3×3 wavefront array constructed on a 4×4 wafer [8.18] having some bad cells. Note the redundant paths used.

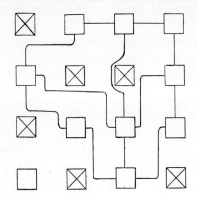

Fig.8.8. An example of reconfigured interconnection in a wafer-scale integrated system

8.6 Conclusion

Image processing is expanding into all technical areas from satellite weather prediction to X-ray medical diagnosis. New applications may be constrained only by the cost and time requirements for the massive amount of computation needed for the image-processing algorithms. With the advent of VLSI and plummeting hardware costs, parallelism is a viable alternative to the escalating computational load. New architectures must be developed to coordinate these concurrent activities and take advantage of technological advances. Finally, to get the most benefit from these new architectures and be cost effective, they must be easily accessible to the users, i.e., easily programmable.

Acknowledgments. The authors wish to thank their colleagues at the University of Southern California for their indispensable contribution and assistance to this research work.

References

8.1 C. Mead, L. Conway: *Introduction to VLSI Systems* (Addison-Wesley, Reading, MA 1980)
8.2 S.Y. Kung, J. Annevelink: VLSI Design for Massively Parallel Signal Processors. Microprocessors and Microsystems (1983). Special Issue on Signal Processing Devices
8.3 H.T. Kung: Why Systolic Architectures? Computer **15**, 1 (1982)
8.4 S.Y. Kung, K.S. Arun, R.J. Gal-Ezer, D.V. Bhaskar Rao: Wavefront Array Processor: Language, Architecture, and Applications. Special Issue on Parallel and Distributed Computers. IEEE Trans. C-**31**, 1054-1066 (1982)
8.5 S.H. Unger: A Computer Oriented Toward Spatial Problems. Proc. IRE **46**, 1744-1750 (1958)
8.6 A. Rosenfeld: Parallel Image Processing Using Cellular Arrays. Computer **16**, 14-20 (1983)

8.7 S.Y. Kung: VLSI Array Processor for Signal Processing. Conf. on Advanced Research in Integrated Cricuits. MIT, Cambridge, MA (1980)
8.8 L.S. Siegel: Pipelined Array Processors. Presented in IEEE L'Aquilla Workshop on Signal Processing (1983)
8.9 H.T. Kung, C.E. Leiserson: Systolic Arrays (for VLSI). Sparse Matrix Symposium, SIAM , 256-282 (1978)
8.10 W.M. Gentleman, H.T. Kung: Matrix Triangularization by Systolic Array, in *Real-Time Signal Processing IV*. SPIE Proc. **298** (1981)
8.11 S.Y. Kung: From Transversal Filter to VLSI Wavefront Array. Proc. Intern. Conf. on VLSI, IFIP, Trondheim, Norway (1983)
8.12 S.Y. Kung, R.J. Gal-Ezer: Synchronous vs. Asynchronous Computation in VLSI Array Processors, SPIE Conf. (1982)
8.13 G.W. Stewart: *Introduction to Matrix Computations* (Academic, New York 1973)
8.14 A. Rosenfeld, A.C. Kak: *Digital Picture Processing* (Academic, New York 1982)
8.15 L.S. Siegel, H.J. Siegel, A.E. Feather: Parallel Processing Approaches to Image Correlation. IEEE Trans. C-**31**, 208-218 (1982)
8.16 S.Y. Kung, Y.H. Hu: A Highly Concurrent Algorithm and Pipelined Architecture for Solving Toeplitz Systems. IEEE Trans. ASSP-**31**, 1 (1983)
8.17 S.Y. Kung: VLSI Array Processors for Signal Processing, Proc. Arab. School Seminars on Modern Signal Processing, Damascus (1983)
8.18 F.T. Leighton, C.E. Leiserson: Wafer-Scale Integration of Systolic Arrays. Proc. IEEE 23rd Annual Symp. on Foundations of Computer Sci., Chicago (1982), pp.297-311

9. Curve Detection in VLSI

M.J. Clarke and C.R. Dyer

Four VLSI designs for a line and curve detection chip are presented. Each method is based on Montanari's dynamic programming algorithm [9.1]. The output of each system is the sequence of coordinates or chain code of the optimal curve(s) detected. One method uses a systolic array of size proportional to the number of rows in the image to detect the optimal curve. To detect a curve of length m requires m passes through the systolic array, taking time proportional to m times the number of columns in the image. A second systolic array design consists of a linearly tapering pipeline of cells. The length of the pipeline is equal to m and the total number of cells in the pipeline is proportional to m^3. Time proportional to the image area is required during a single pass through the image. The third systolic approach uses an array of cells of size proportional to m times the number of rows in the image. Only one pass through the image is required to detect the curve, taking time proportional to the number of image columns. Details are given for a VLSI chip design using the third approach. The fourth method describes an algorithm for an SIMD array machine.

Each of the systolic designs incorporates parallel processing and pipelining using local and regular connections between adjacent processing cells. Each cell is very simple, containing only a small amount of circuitry for performing parallel additions and subtractions.

9.1 Background

Numerous methods have been described for line and curve detection including local linking methods and global methods such as heuristic search, Hough transforms, and dynamic programming. Local methods suffer from a lack of global constraints on the shape of the curve being detected. Heuristic search techniques are inherently sequential and do not permit simultaneous search for multiple paths. Hough transforms are parallel, but require one-to-many mappings to a high-dimensional parameter space.

Dynamic programming has not been widely used in image processing partly because this technique involves excessive computational and storage requirements on a conventional general-purpose computer. In this paper we describe four methods which are appropriate for VLSI implementation and consequently make dynamic programming applications in image processing more cost effective. Specifically, we consider the problem of detecting an optimal smooth curve of length m in an N by M image based on the figure of merit (FOM) function defined by *Montanari* [9.1]. The designs outlined below make extensive use of pipelining and parallel processing (with local and regular connections between adjacent cells) for computing the cost of an optimal curve and detecting the pixels comprising this curve.

A VLSI implementation of dynamic programming for finding an optimal parenthesization of a one-dimensional string has been proposed previously [9.2]. This method, however, only computes the cost of the minimum cost parenthesization, and does not detect it.

In Sect.9.2 we briefly review Montanari's algorithm. Section 9.3 outlines three systolic array approaches to implementing this algorithm. Section 9.4 presents a VLSI design layout for implementing the third systolic method. Section 9.5 describes an implementation of Montanari's method on an SIMD array machine.

9.2 Montanari's Algorithm

Dynamic programming for curve detection consists of three steps: finding the cost of the optimal curve of length m ending at each pixel (using an m stage procedure), finding the global optimal end point in the image, and locating by backward search the other points on the chosen curve. *Montanari* [9.1] defined a simple FOM function for an 8-connected curve based on curvature and gray level. The curve must satisfy a curvature constraint requiring each subsequence of three pixels, z_k, z_{k+1}, z_{k+2}, to change direction by no more than forty-five degrees. Thus for a given pair of pixels, z_{k+1}, z_{k+2}, there are three possible predecessor pixels which are candidates for z_k. The FOM function is given by:

$$g(z_{k+1}, z_{k+2}) = \max_{z_k}[g(z_k, z_{k+1}) + f(z_k) - c] ,$$

where $g(z_k, z_{k+1})$ is the maximum FOM for the pair z_k, z_{k+1} that was calculated at the previous stage, $f(z_k)$ is the gray level at pixel z_k, and c is equal to 1 if the curvature is nonzero and 0 otherwise.

An m-2 stage local, parallel procedure is used to find the maximum FOM for each 8-adjacent pair of pixels, assuming at the kth stage that they are the last two points on an optimal curve of length k+2. At the end of each stage the optimal neighbor (z_k) of each pair is stored. The last two stages are defined specially using a pixel and its eight neighbors. Refer to *Montanari* [9.1] for details. After m stages, the pixel with the largest FOM is the last point on the optimal curve. From this pixel, we then perform a backward chaining search for the remaining points on the desired curve using the "best predecessor" values which were stored at the end of each stage.

9.3 Systolic Architectures

9.3.1 Method 1: Column Processing

In this section we describe one possible systolic design based on parallel processing one column of the image at a time using a single column of cells [9.3]. One cell (consisting of at most three basic cells) per row is used to compute the current maximum FOMs associated with the current pair. Input timing and the staggering of cells by rows is similar to the 3-by-3 kernel cells used in Kung's systolic convolution chip [9.4]. Image columns are processed from left to right in unit time in a synchronous manner.

The FOM for each column of pixel pairs is computed from left to right across the image in the usual systolic manner. Each cell performs computations for all pairs of pixels having the same chain code direction and occurring in the same image row of pairs. Each cell contains three basic cells for receiving image gray levels as input. The spatial organization of basic cells is configured so as to implement the curvature constraint relationship.

Each cell performs in parallel a separate computation in each basic cell (i.e., neighbor, representing an input pair z_k, z_{k+1} for the FOM from the previous stage) and then computes the maximum FOM from these values (three input values for each nonborder pair). Each basic cell is assumed to contain the hardware necessary to implement this computation in unit time. The FOMs that are input to the column of cells at the first stage are initialized to 0.

Each cell outputs the coordinate z_k, where k is the current stage, associated with the maximum FOM and this coordinate is stored in the host's memory associated with the current stage. The pair z_{k+1}, z_{k+2} has an address in memory for the current stage that contains the coordinate z_k associated with the maximum FOM for each pair z_{k+1}, z_{k+2} in the image. Storage of these coordinates at each stage is necessary in order to make the final determination of all those points on the optimal curve (during the backward search phase). Eight arrays of coordinates (one array for each of the eight chain code directions, or pixel pairs) are output and stored at each stage.

Each cell is connected to three other cells (except at borders) with appropriate delays to ensure that the maximum FOM for the pair z_k, z_{k+1} from the previous stage enters the appropriate basic cell at the time required for computing the pair z_{k+1}, z_{k+2}. The required delay is the difference in time between the computation of the pair z_{k+1}, z_{k+2} at the current stage and the computation of the pair z_k, z_{k+1} at the previous stage.

Each stage requires six clock cycles to complete the computations for all columns of pairs in a 4 by 4 image, for example. Each of the eight sets of cells in the column requires control circuity to ensure that the computations of the FOMs begin and end at the correct times. In addition, certain pairs on the border of the image require that their FOMs be set to a value representing $-\infty$ at certain clock cycles. In this way pairs are excluded as possible points on an optimal curve when the curve would be forced to extend outside the image.

Since the last two stages, m-1 and m, are treated differently by the algorithm, either they can be computed separately by the host or two additional stages could be added. The FOMs for the penultimate stage are computed from a pixel and its eight neighbors' values. The last stage requires only the previous stage's FOM at a pixel and its gray level.

Two points in the image will now have the same maximum FOM—the two end points on the optimal curve. One of these points can be found using a systolic comparator array. In case of ties, one end point of one curve is chosen arbitrarily. The coordinates of the end point are used to index into the memory for Stage m-1 to find the penultimate point. The stored maximum FOM directions are then used to detect sequentially the position of each predecessor point on the optimal curve.

The number of interconnections required between cells is minimal in the design described above—approximately 24N interconnections are required for an N by M image. The total number of cells required is 10N-8. The time required to compute the maximum FOM for a curve of length m is O(mM) by a column composed of sets of 8-direction "pair" cells, and O(mNM) host memory is also required.

9.3.2 Method 2: A Pixel Pipeline

Another possible design is to detect the optimal length m curve using an m-stage pipeline [9.3]. Since each pixel's final FOM is effectively a function of the 2m-1 by 2m-1 neighborhood centered around the given pixel, the pipeline is linearly tapering and contains a total number of cells equal to $(2m-1)^2 + (2m-2)^2 + \ldots + 5^2 + 3^2 = O(m^3)$. In order to implement this, the 2m-1 by 2m-1 pixels surrounding each pixel must be input successively using a shift register of size (2m-1)M. (Each pixel is part of neighborhoods for pixels in 2m-1 rows.)

Since at each time step a new 2m-1 by 2m-1 window enters the pipe, the time to compute the maximum FOMs for all pixels in an N by M image is O(NM). Furthermore, computing the global maximum FOM requires just a small amount of additional circuitry at the end of the pipe for comparing each successive pair's final FOM with the current maximum FOM value.

Finding the pixels on the optimal path given the maximum FOM is done as follows. Add circuitry and memory for computing at each cell not only the maximum FOM for every pair associated with a pixel, but also the chain code direction of the best curve ending at the given pixel. These chain code values are output sequentially immediately following their associated end point's FOM (details given below). Thus at any time m curves are at some stage of moving through the pipe towards output at the apex cell. More specifically, at each time step the apex cell stores a chain code link for each of m different pairs' optimal curves. The host stores the current maximum FOM over all possible end points (pixels) seen so far and its associated curve's chain code. At each step the host compares the current best FOM with the (single) FOM value being output at the apex. If the new FOM is greater than the old FOM, then the new value is saved by the host and at subsequent steps its associated chain code is read and stored by the host (replacing the previous curve's chain code).

To accomplish this, each cell at stage k maintains a queue of size 24 · (m-k) bits to store the last m-k maximum pair directions (eight per stage, three bits per chain code direction) for the necessary delay time before each is output. Thus while the first stage computes the FOMs for the set of candidate initial pixels on the optimal curve, these values cannot begin shifting toward the apex until the final pixel's FOM has been computed and all of the successive pixel's on the optimal curve have been found. Only the pixels on the optimal curve shift to the apex; the correct chain code links are selected by a stage's successor stage once its correct link has been

selected. In addition, each cell at stage k stores up to k-1 chain code links
[3(k-1) bits], one for each piece of the k-1 curves which are at some stage of
being output from each of the previous k-1 stages.

If, instead, the k best curves or all curves above a given threshold are
desired, then the host's procedure for curve selection is easily modified to
handle either of these modifications (without affecting the processing in the
pipeline). Finally, if curves of length p > m are desired, then p/m passes
can be used to detect (nonoptimal) curves.

9.3.3 Method 3: A Column Pipeline

In this section we describe a systolic design [9.5] based on parallel processing several columns of an image, each (identical) column of cells performing one stage of processing in Montanari's algorithm. One pair cell (consisting of one to three basic cells) per row is used to compute the current maximum FOMs associated with the current pixels. Input timing and the staggering of cells by rows at each stage is similar to the 3-by-3 kernel cells used in Kung's convolution chip [9.4]. Image columns are processed from left to right in unit time synchronously (Fig.9.1). The FOM for each column of pixel pairs is computed from left to right across the image in the usual systolic manner. Each cell performs computations for all pairs of pixels having the same chain code direction and occurring in the same image row of pairs (links having the same row coordinate for both initial components of each pair); a total of less than 8NM FOM computations are performed at each of the m stages.

Each cell (also referred to as a "pair" cell or a kernel cell) contains three basic cells for receiving the necessary image gray levels (except pair cells at the border of the image which contain only one or two basic cells). Pair cells associated with chain code directions 2 and 6 have no provisions for receiving gray levels from adjacent rows. (We will associate chain code direction k with the vector at angle 45k degrees counterclockwise from the positive x axis.)

At each row in each stage there are a total of eight sets of pair cells — one set for each chain code direction. Each pair cell is connected to up to three pair cells at the next stage. Each cell performs in parallel a separate computation in each basic cell (i.e., neighbor, representing an input pair z_k, z_{k+1} for the FOM from the previous stage). Next, each cell computes the maximum FOM from these values (three values for each nonborder pair). Each basic cell is assumed to contain sufficient hardware to implement this computation in unit time. Initial FOM values equal to 0 are input to the first stage.

Each cell outputs the coordinate z_k, where k is the current stage, associated with the maximum FOM, and also stores this coordinate in the cell's local memory associated with the current stage. The pair z_{k+1}, z_{k+2} has an address in memory for the current stage k that contains the coordinate z_k associated with the direction of the maximum FOM for the given pair. Storage of each of these coordinates at each stage is necessary in order to make the final determination of which pixels are on the optimal curve (during the backward chaining phase). Eight arrays of coordinates (one for each chain code direction) are output and stored at each stage.

Each cell is connected to three other cells (except at borders) with appropriate delay lines to ensure that the maximum FOM for the pair z_k, z_{k+1} from the previous stage enters the appropriate basic cell at the time required for computing the pair z_{k+1}, z_{k+2} at the current stage. The required delay is the difference in time between the computation of the pair z_{k+1}, z_{k+2} at the current stage and the computation of the pair z_k, z_{k+1} at the previous stage, plus two time units (Figs.9.2,3). Each image gray level shifts left in a row in three clock cycles as follows: through three basic

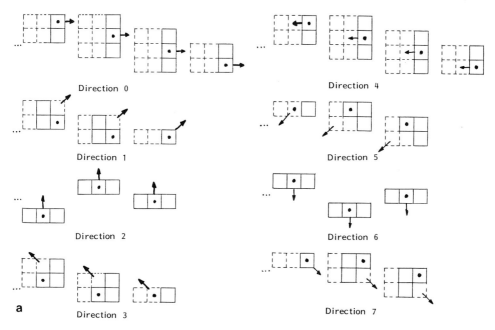

Fig.9.1a. Sets of pair cells needed to compute FOMs for the eight chain code directions

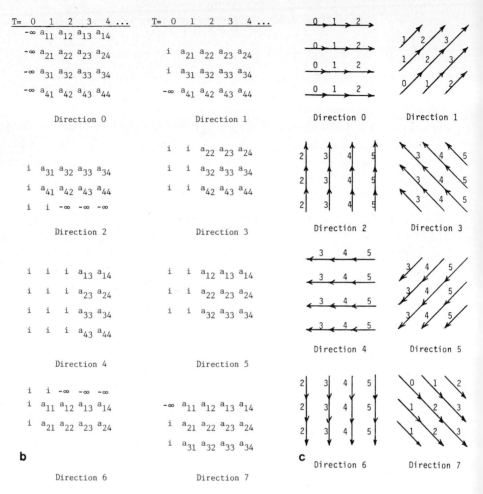

Fig.9.1b. Values stored in the first column of basic cells in Stage 1 at times $T = 0, \ldots, 4$. $a_{i,j}$ is the gray level of pixel (i,j) in the image. (c) Times at which the maximum FOM is computed for each possible pair in a 4-by-4 image

cells in pair cells having chain code directions 2 and 6, through one basic cell and two delay cells for directions 0 and 4, and a combination of three basic and delay cells for directions 1, 3, 5 and 7.

Figure 9.1a shows the FOM computations that are performed for each of the eight chain code directions for links z_{k+1}, z_{k+2}. Within each set of pair cells for a given chain code direction there is a pair cell corresponding to a row in the image; a pair cell computes the maximum FOM for all links z_{k+1}, z_{k+2} that have the same row coordinate for z_{k+1}. Pair cells are denoted

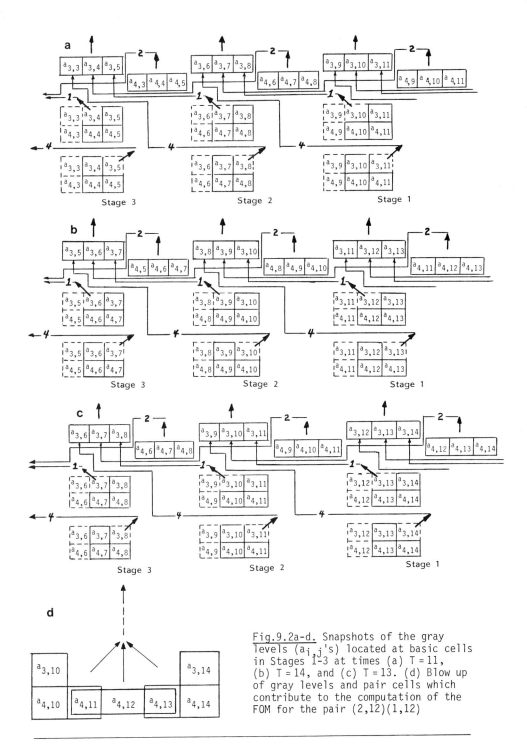

Fig.9.2a-d. Snapshots of the gray levels ($a_{i,j}$'s) located at basic cells in Stages 1-3 at times (a) $T=11$, (b) $T=14$, and (c) $T=13$. (d) Blow up of gray levels and pair cells which contribute to the computation of the FOM for the pair (2,12)(1,12)

165

Fig.9.3. Table of the required interstage delay times for each possible triple of pixels, z_k, z_{k+1}, z_{k+2}. Row numbers refer to the relative row coordinate of z_{k+1}

Direction 0 Interstage Delays			
	Predecessor Direction		
	7	0	1
Row 1	-	3	2
Row 2	3	3	2
Row 3	2	3	3
Row 4	2	3	-

Direction 1 Interstage Delays			
	Predecessor Direction		
	0	1	2
Row 1	4	3	1
Row 2	4	4	1
Row 3	3	-	-

Direction 2 Interstage Delays			
	Predecessor Direction		
	1	2	3
Row 1	4	2	3
Row 2	5	2	3

Direction 3 Interstage Delays			
	Predecessor Direction		
	4	3	2
Row 1	1	1	2
Row 2	1	1	2
Row 3	1	-	-

Direction 4 Interstage Delays			
	Predecessor Direction		
	5	4	3
Row 1	-	1	1
Row 2	1	1	1
Row 3	1	1	1
Row 4	1	1	-

Direction 5 Interstage Delays			
	Predecessor Direction		
	6	5	4
Row 1	-	-	1
Row 2	2	1	1
Row 3	2	1	1

Direction 6 Interstage Delays			
	Predecessor Direction		
	7	6	5
Row 2	5	2	1
Row 3	4	2	1

Direction 7 Interstage Delays			
	Predecessor Direction		
	6	7	0
Row 1	-	-	3
Row 2	1	4	4
Row 3	1	3	4

in the figure by arrows; solid boxes indicate basic cells and dotted boxes indicate unit delay cells which hold image gray levels. Delay cells are necessary in order to guarantee that each gray level is shifted exactly three times (through basic or delay cells) so that a column of image gray levels propagates through each stage at the same rate.

Pair cells having the same chain code direction compute their FOMs for specific pairs z_{k+1}, z_{k+2} in the image with the following spatial organization: pair cells with the same chain code direction at each succeeding stage compute the FOMs of every third pair in the image. A pair at the next stage has column coordinates that are three less than the column coordinates of a pair at the current stage. Two stages of processors are shown in Fig.9.1a. Interconnections between stages for inputting FOMs from the previous stage are omitted in the figure. Pair cells are laid out horizontally in order to improve readability.

The basic cells have a specific row position which is maintained throughout all stages for receiving the gray levels in the corresponding image row. Interconnections for shifting gray levels in a given image row from the leftmost basic cell in one stage to the rightmost basic cell in the next stage for all pair cells with the same chain code direction are shown with dotted lines in Fig.9.1a. The leftmost column of the image enters Stage 1 first. At each clock cycle the image shifts left to the next basic cell or delay cell. In Fig.9.1b T = k indicates that image column k currently occupies the rightmost basic cell in Stage 1. At time T = 0 the indicated column of "don't care" values (i's) has shifted into the rightmost column of basic cells.

At a given clock cycle the FOMs for an image column consisting of eight sets of pairs (z_{k+1}, z_{k+2}) are computed at each stage. When gray levels are in basic cells, indicated by dots in the figure, the associated pair cell performs a FOM computation for each gray level (in parallel) and outputs the maximum FOM. The functional significance of the "don't care" values (i), $-\infty$, and basic cells designated by dots is illustrated by the FOM computation for the lower-left border pair with chain code direction 2 (DIR-2). At time T = 2 the sequence i, $-\infty$, $-\infty$ occupies the basic cells in a pair cell at Stage 1 (the lowermost pair cell of the three associated with DIR-2 in Fig.9.1a). At T = 2, the first FOM computation and maximization are performed by this pair cell — the FOM for this pair is set to $-\infty$ (see *Montanari* [9.1] for details on the use of this parameter). In Fig.9.1c the clock cycle at which the maximum FOM is computed for each possible pair in an image window consisting of four rows is indicated above each pair (arrows) for Stage 1. The times of the FOM computation for each pair at any succeeding stage k can be determined by adding the number 3(k-1) to each indicated time (shown above each image pair in Fig.9.1c).

Figure 9.2 shows the image gray levels which are located in basic cells of pair cells and the interconnections (dotted lines) used for transmitting FOMs to two pair cells (at Stages 1 and 2) having the same chain code direction. The spatial distribution of gray levels in basic cells is shown for clock cycles T = 11, 13, and 14 to illustrate the spatial and temporal properties of the pipeline. Figure 9.2d shows the spatial orientation of pairs for the previous stage with coordinates (3,11)(2,12), (3,12)(2,12) and (3,13)(2,12) (shown with solid lines). The three associated maximum FOMs of these predecessor pairs are input, after appropriate delays, to a pair cell at the current stage corresponding to the pair (2,12)(1,12). The gray levels of pixels (4,11), (4,12) and (4,13) are now in the correct basic cell positions at time T = 16 for the FOM computation of the DIR-2 pair at coordinates (2,12)(1,12). A delay of two clock cycles is required before inputting the FOM (computed at Stage 1) from the predecessor pair DIR-2 at coordinates (3,12)(2,12) at Stage 2. From Figs.2a,c it is readily verified that the required delay times for predecessor pairs (3,11)(2,12) and (3,13)(2,12) are 16-11-1 = 4 and 16-14-1 = 1, respectively.

The interstage delays required before inputting the FOMs from predecessor pairs z_k, z_{k+1} to the eight sets of pair cells at the current stage are shown in Fig.9.3 for a window consisting of four image rows. Rows within each set refer to the row coordinate of z_{k+1} for the pair z_{k+1}, z_{k+2} that receives a FOM from three possible pairs z_k, z_{k+1}.

At the last stage, the final FOMs for a column of image pairs are output (except for the final FOMs for chain code directions 1 and 7 which are output from the next to last column). At the last stage one of three possible pairs z_{k+1}, z_{k+2} (associated with the optimal curve ending at pixel z_k) has the maximum FOM. A systolic comparator at the output of the last stage is used to detect this pixel's maximum FOM. The end point z_m of the global optimal curve is recovered from the last stage's memory at a storage location with address z_{m+1}, z_{m+2}. The coordinates of z_m, z_{m+1} are then used to index into Stage m-1's memory to find the penultimate point on the optimal curve. Similarly, the stored maximum FOM directions are used to detect sequentially the position of each predecessor point on the optimal curve.

In summary, approximately 24Nm interconnections are required for inputting the FOMs for an N by M image. The total number of cells required is m(8N-6). The time required to compute the maximum FOM for a curve of length m is O(M+m). O(NMm) memory is also required. O(m) time is used to output the optimal curve.

In order to detect curves of unknown length, it is necessary to normalize the FOM at each stage and store the normalized maximum FOM. To implement this, pair cells could be preloaded with constant multipliers having magnitudes equal to the reciprocal of their stage number. The FOMs that are transmitted to succeeding stages are not normalized. Additional circuitry is also necessary for detecting the stage at which the normalized FOM exceeds a given threshold, indicating the end point of a "short" curve. Curves which are longer than the given number of stages must be detected in pieces and linked together (nonoptimally) later.

9.4 A VLSI Design of the Column Pipeline Chip

In this section we describe a possible VLSI design of the third systolic method presented in Sect.9.3.3. Figure 9.4 shows a VLSI chip design and pin configuration for implementing one stage using this approach. Three image rows can be processed by this chip; m of these chips connected in series

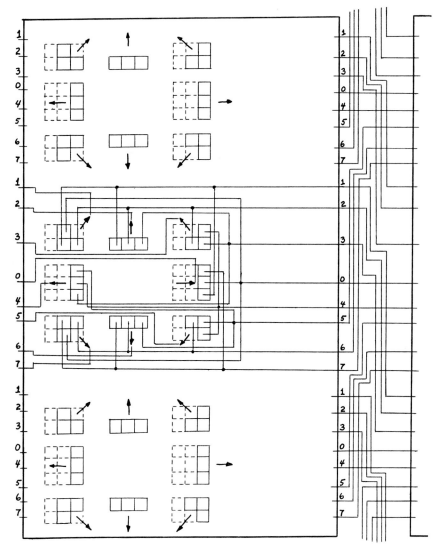

Fig.9.4. Layout of pair cells for a chip which processes three image rows and a single stage of the systolic column pipeline method

would be required to detect curves of length m. Additional chips must also be connected above and below to allow processing of a complete image in a single pass — a column of M/3 chips to process images with up to M rows. Five pins are required at the right side of the chip for inputting gray levels and five pins at the left side for shifting gray levels to the next stage. The horizontal and vertical lines connected to dots in the rightmost basic cells (dotted squares) distribute gray levels to the pair cells. The pair cell configuration shown in the figure was designed to minimize the wire distances for distributing gray levels and to provide simple and regular wire connections for inputting the FOMs from the three neighboring chips at the previous stage. The objective was achieved by allocating each of three sets of processors on the chip to the processing of a row of image pairs (pairs z_{k+1}, z_{k+2} having the same row coordinate for z_{k+1}).

Shown at the extreme right in the figure are the numbered pin connections of the adjacent chip at which FOMs are output from the preceding stage. Each numbered pin is connected to a pair cell in the preceding stage for the purpose of inputting the FOM to the next stage via off-chip connections. Ignoring border pairs, three pair cells at the next stage have a basic cell that receives a FOM input.

The FOM computed by each pair cell on the chip is transmitted to the correspondingly numbered pins on the left side of the chip. Not shown in the figure is the external wiring pattern on the left side of the chip which transmits the FOM values to the next stage. Figure 9.4 corresponds, then, to a chip in the last stage. At the last stage the FOM pins on the left side of the chip are connected to a comparator for determining the global maximum FOM.

In Fig.9.4 three gray levels are input serially in the following manner: three adjacent rows are input at the central pins on the right side of the chip (after having shifted through the chips for the preceding stages). If a chip is connected to other chips above and below it (for processing additional image rows), then gray levels for the row above the current chip's top row and below its bottom row are input at the upper-right and lower-right sets of pins, respectively. Basic cells of certain pair cells (shaded squares) are inactivated by control inputs. The inactive cells shown in Fig.9.4 are those needed for an image containing three rows.

As an example, consider the processing of a 60 by M image to detect the optimal curve of length thirty. An array of 20 by 30 chips like the one shown in Fig.9.4 is needed in this case. Assuming gray levels are quantized to sixty-four levels, the FOM computation at each basic cell requires an eleven-bit serial adder and a parallel subtractor (for finding the maximum FOM at a pair cell having three basic cells).

Additional circuitry is required to detect and output the chain code direction of the best pair. In each pair cell a counter is incremented by one after each clock cycle to keep track of the column coordinate of the pixel z_{k+1} involved in the current FOM computation. To process a 1000 by 1000 image 3K bits are required for each pair cell's local memory.

A threshold device at each FOM output pin at the last stage could be used to initiate the sequential outputting of all coordinates on the optimal curve. The coordinate stored in the last stage's memory is output first. Each coordinate is transmitted serially via FOM connections through m-k chips and is output on the pin where the final FOM was detected (one of pins 0-7 on chips in the last stage). The chain code direction that is associated with the pair z_k, z_{k+1} selectes the corresponding line (three demultiplexors per chip are required) of the eight possible FOM lines externally connected to the preceding stage. The coordinate z_{k+1} is then transmitted to the pair cell having the same chain code direction in one of the three neighboring chips to the right. At Stage k-1 the coordinate z_{k+1} is computed using the chain code direction associated with the pair cell at Stage k-1 and the coordinate z_{k+1} from Stage k. The next point on the curve, z_k, at Stage k-1 is recovered from the memory location associated with the designated pair at a location corresponding to the address of z_{k+1}; this coordinate is then passed right to the next stage. This recovery process is repeated for each preceding stage until the entire curve is extracted.

9.5 SIMD Array Algorithm

An alternative method for implementing Montanari's algorithm would be to use an SIMD array machine such as CLIP [9.6]. Assuming one processor per pixel, each processor computes its pixel's associated pair's FOMs in O(m) time. Since there are approximately eight times more pairs that require a FOM computation than there are processors, eight sequential sets of operations, one set for each pair's chain code direction, are required for computing the given pixel's new FOMs at each of stages 1 to m-2. Propagation commands load the image gray levels and the FOMs computed at the previous stage for the pairs z_k, z_{k+1} into the processor assigned to pixel z_{k+1}. The curvature penalty is sent by the array controller for the current propagation input direction and the current chain code direction of the pairs z_{k+1}, z_{k+2}. The above instructions are performed once for each of the three possible inputs (z_k, z_{k+1}) satisfying the curvature constraint followed by an instruction that computes the maximum FOM of these three values.

The coordinate z_k associated with the maximum FOM and the FOM itself are stored by the processor assigned to the pixel z_{k+1} for the pair z_{k+1}, z_{k+2}. All the instructions just described are executed by the processors assigned to z_{k+1} eight times for all eight chain code directions of pairs z_{k+1}, z_{k+2}. These eight coordinates and their associated FOMs are then passed on to the next stage. Only the maximum FOM and associated direction need to be permanently stored by the processor for subsequent use during the backward chaining phase.

The second to last stage's computations require propagation commands to load each processor with the FOMs associated with each of its eight neighbors. The coordinate (i.e., chain code direction) of the neighbor with the maximum FOM is stored at each processor. At the last stage, each processor adds the maximum FOM just calculated to the gray level associated with the given processor.

The location of the optimal curve can be output to the host through a designated processor, say the upper-left corner one, as follows. The final FOMs from the mth stage are shifted leftwards and then upwards to the upper-left corner processor which computes the maximum FOM in $O(M+N)$ steps. This value is then broadcast back to each pixel. The two pixels whose final FOMs are equal to the broadcast value are the two end points of the optimal curve. Each of these two end points then (independently) shifts its coordinates to the output processor, with the remaining points on this curve "snaking" through the array, each following the point ahead of it. Total time to detect and output the optimal curve is $O(m+M+N)$.

9.6 Concluding Remarks

In this paper we have outlined four possible implementations of Montanari's algorithm for detecting curves of length m. Each approach utilizes parallel processing by a large number of simple, identical processing elements, making each appropriate for VLSI implementation. The systolic column-pipeline approach is the fastest, taking $O(M+m)$ time. The SIMD array algorithm is comparable in speed [$O(m+M+N)$], but requires the largest number of processors [$O(N+M)$]. The systolic column-processor requires $O(N)$ processors and $O(mM)$ time (plus additional I/O time for storing and retrieving intermediate FOMs in host memory during the m pass procedure). The systolic pixel-pipeline approach requires only $O(m^3)$ processors, but $O(NM)$ time (only a single pass through the image, however).

Of course, if the curve is known to be present in a window of the image, then only the relevant columns and rows need to be input and processed. Both storage and computation time can be reduced by searching for p segments of size m/p and then performing a linking operation to obtain the complete curve. Of course, this method is not guaranteed to be optimal.

References

9.1 U. Montanari: Comm. ACM **14**, 335-345 (1971)
9.2 L.J. Guibas, H.T. Kung, C.D. Thompson: "Direct VLSI Implementation of Combinatorial Algorithms", in Proc. Caltech Conf. on Very Large Scale Integration: Architecture, Design and Fabrication (1979) pp.509-525
9.3 C.R. Dyer, M.J. Clarke: "Optimal Curve Detection in VLSI", in Proc. IEEE Conf. Computer Vision and Pattern Recognition (1983) pp.161-162
9.4 H.T. Kung, S.W. Song: "A Systolic 2-D Convolution Chip", in *Multicomputers and Image Processing*, ed. by K. Preston and L. Uhr (Academic, New York 1982) pp.373-384
9.5 M.J. Clarke, C.R. Dyer: "Systolic Array for a Dynamic Programming Application", in Proc. 12th Workshop on Applied Imagery Pattern Recognition (1983)
9.6 T.J. Fountain: "Towards CLIP6—an Extra Dimension", in Proc. IEEE Workshop on Computer Architecture for Pattern Analysis and Image Database Management (1981) pp.25-30

10. VLSI Implementation of Cellular Logic Processors

K. Preston, Jr., A.G. Kaufman, and C. Sobey

Work on cellular logic computers was first inspired by the writings of *von Neumann* [10.1] on cellular automata. Now such machines are in use in hundreds of laboratories worldwide primarily in the field of two-dimensional array processing for image analysis. It took many years to reduce the ideas of von Neumann to practice in hardware, because until recently, it was impractical to build computing machines consisting of millions of devices. Only with the development of large-scale integrated circuitry (LSI) in the 1970s has this feat been accomplished. The major impact of LSI, and now VLSI, has been in the field of full-array automata as represented by the Cellular Logic Image Processors (CLIP) in operation since 1980 at University College London [10.2] and the MPP (Massively Parallel Processor) of the United States National Aeronautics and Space Administration (fabricated by the Goodyear Aerospace Corporation) operational at NASA Goddard since mid-1983 [10.3]. These full-array machines are significant in that their processing elements (one-bit microprocessors) are integrated at eight per chip, their operational speeds are equivalent to tens of billions of instructions per second, and they represent the fruition of von Neumann's visionary work in the 1950s. These full-array machines, however, are limited in their array size. The CLIP machine is 96×96 and MPP is 128×128. Unfortunately, many images of interest are 1024×1024 and larger. Also, as the array size has grown, expense has spiraled. The MPP is priced at several million dollars.

10.1 Cellular Logic Processors

An attractive alternative to the full-array cellular automaton is the cellular logic processor, a subarray machine, which emulates the array automaton. Such machines are more economical and not limited to a fixed, small image size. They were first produced in discrete-component form in the early 1960s, beginning with CELLSCAN, the world's first cellular logic machine built by *Preston* [10.4]. This machine and other special-purpose machines emulated the cellular automaton by using one or more high-speed processing elements to operate sequentially on an array of bilevel data. By the 1970s, such machines were in production using a multiplicity of processing elements, e.g., the diff series of machines produced by Coulter Electronics Inc. [10.5] and the cytocomputer series of machines described by *Sternberg* [10.6].

This chapter reviews the evolution and architecture of cellular logic machines and then describes present efforts to produce such machines using LSI/VLSI. The significance of these machines is that they are now widely used both for commercial and research purposes. Despite their inherent limitation to handling bilevel image data, they are also employed in gray-level image processing due to their ability to convert gray-level images to bilevel images by multiple thresholding. After multiple thresholding, these machines carry out cellular logic operations on the resultant data and, finally, generate gray-level output data by arithmetic summation.

10.2 Binary Neighborhood Functions

A bilevel (or "binary") image is one in wich the value of each picture element is represented by a single bit. Such images are, therefore, "black and white". They are processed or modified by use of logical rather than numerical transforms employing Boolean algebra. As mentioned above, gray-level images may be converted to a registered stack of binary images through multithresholding. Then each member of the stack is transformed either individually or in Boolean combination with other members. The results stack is then returned to gray-level format by simply adding all members to produce an array of integers.

Logical processing often has advantages over traditional numerical methods in that multilevel logical transforms followed by arithmetic recombination have certain unique properties. Logical transforms may be considered as filters [10.7]. In this regard, some have the valuable property of being constant phase with zero sidelobes, i.e., passing absolutely no signal beyond their cutoff frequency (Fig.10.1). In turn, this cutoff frequency may be adjusted by performing a number of iterations of the transform selected. Furthermore, logical transforms, when executed as convolution functions using small kernels, may be executed at ultrahigh speed (less than one nanosecond per convolution step) by doing all computation using table lookup and paralleling lookup tables. This is the philosophy behind the machines described in this chapter.

Cellular logic computers are used for the digital computation of two-dimensional and, recently, three- and four-dimensional logical neighborhood

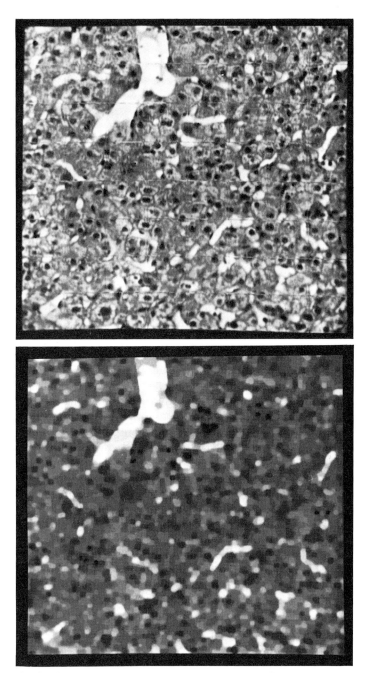

Fig.10.1. Human liver tissue showing both cell nuclei and vessels (top) and the results of multithreshold logical processing (bottom) using a sequence of 3 × 3 logical convolutions followed by arithmetic summation as discussed in [10.7]. (Reprinted with permission of the Institute of Electrical and Electronics Engineers, copyright 1983)

functions in both monochromatic and multichromatic image processing [10.8]. For example, a 3 × 3 convolution function may be calculated by the cellular logic computer and used for nonlinear image enhancement, boundary detection, texture analysis, etc. Furthermore, when information on picture-element to picture-element connectivity is utilized, these transforms are employed in object counting and sizing, shape analysis and skeletonization. An example is given in Fig.10.2.

10.3 CELLSCAN

As mentioned above, the first cellular logic machine was CELLSCAN, built by the Research Engineering Department of the Perkin-Elmer Electro-Optical Division under contract to the University of Rochester using funds provided by the United States Atomic Energy Commission. CELLSCAN was structured to operate in a serial mode using a single processing element operating in parallel on a 3 × 3 subarray. The entire system consisted of a slow-scan television microscope for image input interfaced to the CELLSCAN special-purpose computer. This computer was fabricated in 1961 for Perkin-Elmer by the Navigation Computer Corporation (NAVCOR). Image data was transferred to the computer at a 60 Hz horizontal rate. The frame time was 5 seconds and an image field was produced having 300 horizontal lines and a total of 90,000 image elements. By means of video preprocessing, this 300 × 300 array was sampled and threshold into a binary array of 63 × 300. The data was stored on an endless magnetic tape. Picture element data was read from the tape one element at a time, the value of a new image element was computed, and the result rewritten onto the tape. Thus, the tape served as a 18900-bit delay line. Data was recorded in two complementary tracks for error detection and to permit clock pulses to be generated when either a 1-element or a 0-element appeared under the reading heads.

Image data flowed from the two 60-bit registers which held incoming image element values and then into a single 60-bit "look back" register. A 9-bit register was used to hold the 3 × 3 subarray data used in calculating the output function (Fig.10.3). Control registers were used to perform various counting functions and to indicate both start of line and start of field. Another register was used to count the number of iterations used to operate upon the entire data field. Programming was carried out solely by means of sense switches on the console with only a limited number of image processing operations available. These included (1) an operation wherein groups of contiguous binary 1-elements in the data field were reduced to residues (isolated

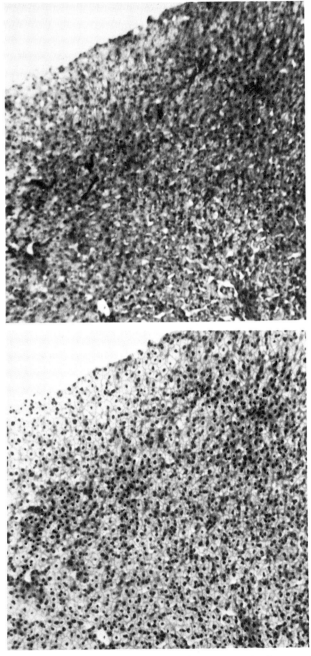

Fig.10.2. By using a complex series of logical transforms after multithresholding the image shown on the top, the location of both cell nuclei and an estimatin of the position of cell borders is obtained (bottom). (Reprinted with permission of the Institute of Electrical and Electronics Engineers, copyright 1983)

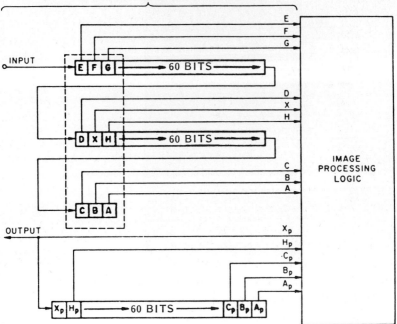

Fig.10.3. Data flow of image information through the CELLSCAN shift registers, the neighborhood register, and the processing element logic. (Reprinted with permission of Auerbach Publishers, Philadelphia, copyright 1973)

1-elements surrounded by an eight-element neighborhood containing nothing but 0-elements); (2) counting of residues; and (3) binary complementation of the image. These functions were carried out by wired circuitry using discrete-component transistor logic. A detailed discussion of the entire CELLSCAN system is given by *Izzo* and *Coles* [10.9].

10.4 The Golay Parallel Pattern Transform

In the mid-1960s, *Golay* [10.10] invented the Golay parallel pattern transform. This invention solved the problem of requiring recursive operations (the "look back" register of CELLSCAN) in order to carry out connectivity-dependent transforms such as skeletonization. Golay's solution was to divide the image data field into three disjoint subfields whose union was the entire field. To take advantage of this new methodology, the construction of a second-generation cellular logic machine was commenced in 1967 by the Optical Group Research Division of Perkin-Elmer, resulting in the completion by 1968 of the Golay logic processor (GLOPR) which incorporated not only subfield

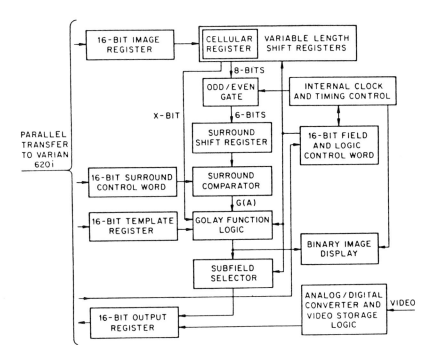

Fig.10.4. Block diagram of the GLOPR cellular logic machine showing the five 16-bit registers used for parallel/serial transfers to the host minicomputer. Two registers hold control words, two are used to transfer the A image data field and the template data field (B or C). Another register is used in transferring either output data or digitized image data from the associated television scanner to the host computer. (Reprinted with permission from the Auerbach Publishers, Philadelphia, copyright 1973)

logic but also hexagonal tessellation. The GLOPR took full advantage of the integrated circuits then available (SSI/MSI), was fully programmable, and could handle multiple data fields in a generalized fashion. Furthermore, GLOPR was built as a peripheral to one of the early minicomputers (Varian 620i) configured with 32 K bytes of main memory and a seven-track magnetic tape (Fig.10.4). The GLOPR was driven by a language called GLOL [10.11] which included several innovations in programming cellular logic machines. Furthermore, GLOL has a data type for arrays, as well as traditional scalars and indexed vectors. Arrays in GLOL are not indexed and are all simultaneously dimensioned by a SET SYSTEM command to be either 128×128, 64×64, or 32×32. It was the first structured language for image processing in that it permitted no labels and outlawed the GOTO statement. Thus all code is entirely in line. Debugging, even of elaborate programs, is extremely simple.

Additionally, GLOPR exhibited innovations in that it was both microcoded and vectored. The microcode consisted of two 16-bit words which were transferred from the minicomputer at the beginning of each image transform. One word used 14 of its 16 bits to indicate the particular combination of Golay primitives to be monitored in the hexagonal tessellation. (Since there are 2^{14} combinations of the 14 hexagonal primitives of Golay, all 14 bits are required.) The other microcode word used two bits to specify the size of the data field, two to point to the subfield, three to give the subfield sequence, and eight to set up the Boolean algebraic portion of the Golay transform. In addition to the two microcode registers, GLOPR also contained two registers to receive input data fields and an output register to hold the contents of the destination data field. (This register could also be used to transfer incoming digitized video information from an associated scanner.) The clock rate of GLOPR was 4 MHz. Eight clock cycles (2 μs) were used to generate the timing sequence for processing each image element. After operating on a group of 16 bits from the data field, GLOPR initiated several I/O transfers adding approximately 50% overhead. Thus 48 ms were required to process the 128×128 data field, making GLOPR approximately 200 times faster than CELLSCAN.

10.5 The diff3-GLOPR

The Golay parallel pattern transform and the GLOPR architecture for the basic cellular logic machine was continued in the diff series of machines initially commercialized by Perkin-Elmer and later transferred to Coulter Electronics Inc. The first machine in this series was called "diff3", from which evolved the diff3-50 and the diff4 [10.5]. The GLOPR computers in all of these machines are identical and are a major advance over the initial research GLOPR produced in 1968. This new GLOPR introduced multiple subarray processors operating in parallel. Instead of a single subarray processor examining one hexagonal neighborhood at a time, the GLOPR of the diff series uses eight parallel processors.

In this new GLOPR the rotating neighborhood shift register of the early research GLOPR was eliminated and elements of the neighborhood were decoded directly by table lookup using semiconductor ROMs. A single ROM with $2^6 = 64$ addresses is used to identify the 14 Golay primitives in the six-element neighborhood. This new GLOPR uses eight parallel ROMs, each generating a one-hot-out-of-14 signal according to the Golay primitive contained in the corresponding input neighborhood. These signals are then compared with the contents of the appropriate microcode control word and with the present value of the central element to determine the new value of the central element. The advantage of this architecture is that the microcode used to specify the primitives for a particular transform consists of a single word whose length is equal to the total number of Golay primitives. The alternative (used in the PHP discussed below) is storage of a complete lookup table consisting of $2^7 = 128$ words of microcode with each word containing a new value of the central element corresponding to the present logical configuration of the neighborhood and the present value of the central element.

The final computational stage of the diff-series GLOPR is performed in a set of eight combination logic circuits. Inputs to these circuits consist of the eight center elements from input image A and the center element of input image B. Output data for the results are placed on a 16-bit bus connecting to the destination image register. As the output data are generated, a set of eight high-speed counters is employed to totalize the bits in the destination image. At the end of a complete Golay transform cycle, the contents of these counters are added to produce the final output count for the transformed image. This count is the measurement data output which is transferred from the GLOPR to the control computer. In the diff3 and the diff3-50, the control computer is the Data General Nova, while in the diff4, the control computer function is performed by a multimicroprocessor system consisting of thirteen Intel 8086 microprocessors.

10.6 The Preston-Herron Processor (PHP)

The next step in this evolutionary chain was the PHP designed by Carnegie-Mellon University and fabricated by the University of Pittsburgh under contract to the Perkin-Elmer Corporation [10.12]. Like GLOPR, the PHP was designed for use as a peripheral to a host minicomputer. It differs from GLOPR in the use of the square tessellation, sixteen lookup tables, and full decoding of the cellular logic transformed by employing 512-position RAMs. Thus, the PHP discards the idea of the Golay primitives and uses lookup tables configured in LSI. These tables employ $2^9 = 512$ addresses. This required the use of sixteen RAMs.

The host computer is used to load the lookup table in all RAMs in parallel, threshold gray-level image data to produce binary data, load sequential portions of this data in the PHP, and unload the binary result image (Fig.10.5). This I/O function is carried out 32 K image elements at a time using a "memory redundancy" feature wherein the image data are loaded into three identical VLSI 2048 × 16 memories where all three contain the same information.

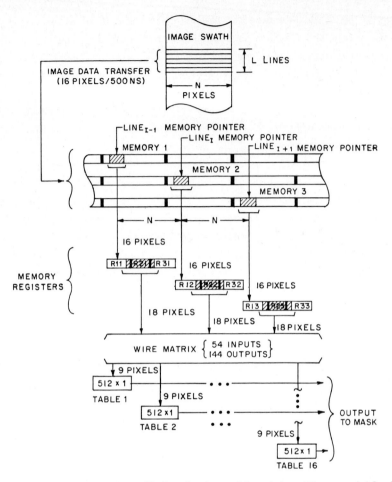

Fig.10.5. In the PHP cellular logic machine data with a variable line length L are received from the binary image store. One image slice, consisting of N lines, is stored redundantly in three identical image memories. Offset addresses previously entered by the host computer are then used to gate the values of 48 picture elements simultaneously to the memory registers. These registers in turn feed 16 lookup tables, whose output is returned to the host computer via other gates whose output is controlled by the contents of the subfield register. (Reprinted with permission of the Institute of Electrical and Electronics Engineers, copyright 1982)

This permits the data to be read from these memories and processed by table lookup as a 18×3 subarray. The result of the table lookup operation is then transferred to the memory of the host computer. Masking is provided at the output of the PHP to permit subfield operations in the square tessellation.

Overall image dimensions are determined not by the PHP but by the host computer. For example, a square image as small as 16×16 or an image slice

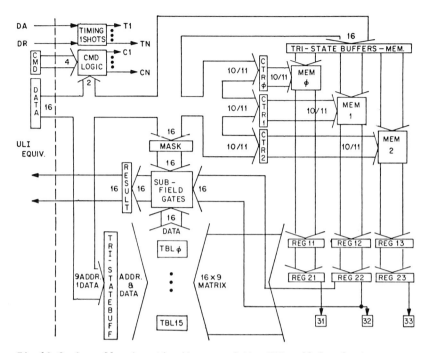

Fig.10.6. Overall schematic diagram of the PHP cellular logic processor showing control and data paths. Image data is entered as pairs of 8-bit bytes in parallel from the host computer into the registers labelled MEM. Offset addresses are 11 bits so that as many as 32768 picture elements may be addressed (16 at a time). Lookup table addresses are 9 bits and are loaded one address at a time. Control is from four command latches whose contents are decoded to provide five major modes of operation. (Reprinted with permission from the Institute of Electrical and Electronics Engineers, copyright 1982)

as large as 3×10912 can be processed. Intermediate sizes can be handled as long as the number of lines are selected modulo 1 and the number of pixels per line selected modulo 16. This permits images to be processed which are essentially infinite in length and 10912 pixels wide. In order to process the section of an image, commands are issued (1) to load the subfield mask; (2) to load the lookup tables; (3) to set the PHP memory address offsets to the corresponding image dimensions; (4) to load all three memories in parallel with identical copies of the image slice to be processed, and (5) to perform the processing sequence transferring data through the memory registers 48 elements at a time, addressing the 16 lookup tables in parallel, and generating a 16-bit masked output, Fig.10.6. At each step in the processing cycle, the 16 elements that are the contents of the output register and their 16 neighborhoods (54 elements in all) are available as sixteen 9-bit individual table addresses (on 144 total wires) to the RAMs. These tables furnish information to the output logic, which, in conjunction with the subfield mask, gates the 16 binary results to the host computer.

The PHP is capable of loading and processing 2 Mbytes (16 million image elements) per second. Thus, an image slice consisting of 32 K 1-bit elements could be processed in about 5 ms (one input cycle plus one output cycle). To this time must be added the overhead time required to enter the tables (equivalent to 1024 bytes), to load the subfield mask, and to load the memory address offsets. Depending on the time-sharing load on the host computer system and other variable factors, it could take as short a time as 100 ms for the PHP to process a 512×512 image. Two PHPs are now in operation; one at the University of Pittsburgh and one at Perkin-Elmer, Danbury, Connecticut.

10.7 The LSI/PHP

The purpose of this section is to describe efforts to transform the PHP design architecture to the requirements of VLSI implementation. This was commenced during 1983 by one of the authors (Kaufman) as part of the VLSI design course at Carnegie-Mellon University under the instruction of Professor Marc Raibert. Since the PHP consists of some 150 (MSI/LSI/VLSI) chips (32 K bits of image slice memory, sixteen 512-bit RAMS, memory registers, control logic, and host computer interface) it was clear that it would be necessary to partition the system. Also an overall configuration was needed which would employ the resulting LSI/PHP chip effectively in a time-sharing environment.

10.7.1 General Design Configuration

The general design configuration selected is shown in Fig.10.7. Initially the LSI/PHP is to be controlled only by a Motorola 68000 microprocessor configured on a standard VM02 board having 128 K PROM for its operating system, editor, and assembler plus 128 K RAM for user programs. A standard RS232 port is to be used to interface the user to the system via a Video Display Terminal (VDT). A set of LSI/PHP chips is to be interfaced to the 68000 via its Versabus. Later in the project it is intended to add a further interface between the 68000 and the mainframe now used with the PHP for image processing. This computer system is based on Perkin-Elmer 3230, a 32-bit minicomputer system configured with several VDTs, a 300 MB image storage disk, a 80 MB systems disk, a Versatec printer-plotter, and a Ramtek full-color display. This system will permit down-loading image data to the 128 K RAM of the 68000 plus additional data providing instructions as to the sequence of

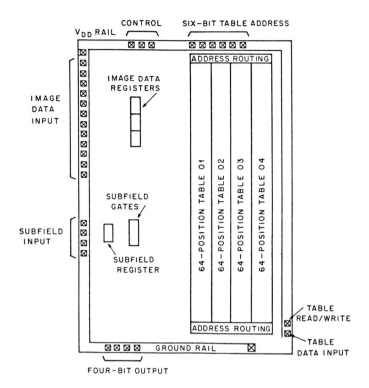

Fig.10.7. Floorplan of the preliminary design of a LSI/PHP consisting of a segment of four lookup tables operating in the hexagonal tessellation. The general design philosophy is similar to the original discrete-component PHP (Fig.10.6)

logical transforms to be performed. The 68000-LSI/PHP subsystem would then assume control and perform all required operations. Finally, this latter system would interrupt the PE3230 and return the resultant image (or images) to the host.

10.7.2 Chip Design Trade-Offs

After a thorough investigation of design trade-offs, it was clear that the hexagonal logical transform was preferable to the logical transform carried out in the square tessellation, because the lookup table for the square tessellation requires 800% more devices than the hexagonal. Using a 500 × 500 mm chip, it was decided to employ approximately 50% of the available area for lookup tables and, further, to limit the table addressing to the six-element neighborhood. Static RAMs rather than dynmaic RAMs were employed. This led to the floorplan shown in Fig.10.8 wherein four 64-position static RAMs are

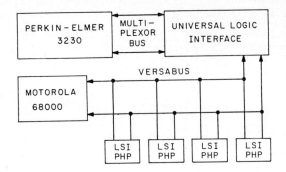

Fig.10.8. Schematic showing the interconnection of four PHP chips via a Motorla 68000 to the Perkin-Elmer 3230 host computer. At present only the Motorola 68000 interface is being implemented

deployed on the chip for this purpose. Because of the large area utilized for lookup tables, it was impossible to provide any significant number of bits for image slice storage. Therefore, the next decision was to include on the chip only the memory registers, the interconnection logic between these registers and the lookup tables, as well as the subfield mask register with its output logic as shown in Fig.10.8.

The remaining area available is presently being reserved for interface logic to the Motorola 68000. Thus, with the present configuration, the 68000 will perform the function of memory data indexing in the process of transferring image data from its main memory registers on the chip. The primary function of the chip is to carry out the logical transform by table lookup and handle subfield masking. Prior to this, the 68000 will pass lookup tables to the chip as well as storing the subfield mask in the mask register. Output data will be placed on the Versabus during the processing cycle and will be transferred by the 68000 either to an accumulator in its main memory or rewritten over the original image input data.

10.7.3 Design Methodology

To design and fabricate the LSI/PHP chip, the ICARUS software package was employed at Carnegie-Mellon using the Xerox Alto VLSI design workstation. The ICARUS package is an interactive CAD/CAM system which permits the user to lay out the masks required to fabricate NMOS integrated circuits. Masks may be generated for diffusion, implantation, epitaxial growth, and metalization as designated by the "stipple" patterns given in Table 10.1. Furthermore, ICARUS allows the user to designate rectangles or "items" generated by drawing horizontal, vertical, and $45°$ lines using the DRAW mode. Each item so defined is stored in six words of computer memory up to a maximum of approximately 300 items. These items and their combinations form the masks. When previously defined items are utilized they are accessed via the

Table 10.1. ICARUS VLSI Graphics Symbols

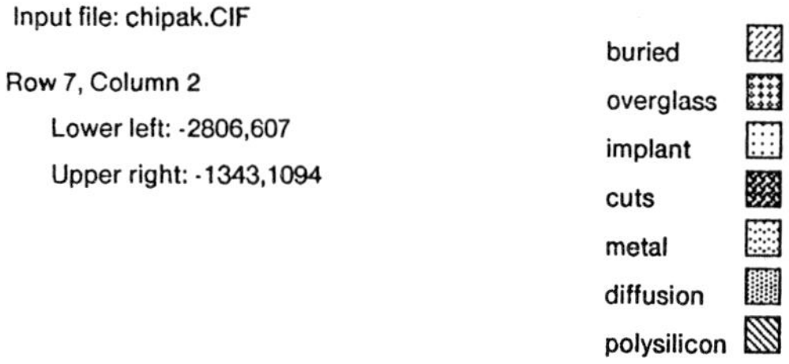

Input file: chipak.CIF
Row 7, Column 2
Lower left: -2806,607
Upper right: -1343,1094
buried
overglass
implant
cuts
metal
diffusion
polysilicon

SELECT mode and positioned by the MOVE sequence. They may be scaled by means of the STRETCH command.

In many cases, a building block (such as the static RAM memory cell shown in Fig.10.9) consists of a multiplicity of items. Such a building block may be stored on disk as a "symbol" using the DEFINE SYMBOL command. This command labels the symbol and the DRAW SYMBOL command is used in constructing the symbol. The complete symbol for the memory cell shown in Fig.10.9 is shown in Fig.10.10. As with items, symbols may be moved by the MOVE command.

The user of ICARUS is provided with two display windows on the display screen. One window displays a major portion of the system under design while another provides a magnified view of a portion of this design. A SCROLL command permits a design that is too large to be displayed to be scrolled through

X_i
X_i
X_{i+1}
X_{i+1}
R/W O
R/W 1

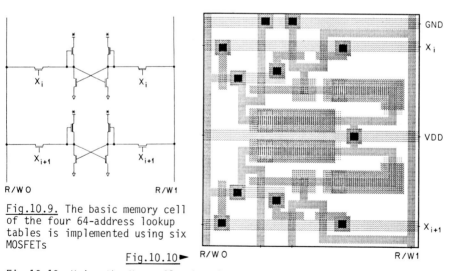

Fig.10.9. The basic memory cell of the four 64-address lookup tables is implemented using six MOSFETs

Fig.10.10►

GND
X_i
VDD
X_{i+1}
R/W O
R/W 1

Fig.10.10. Using the Xerox Alto-based ICARUS system a "symbol" design shown here may be drawn, labeled, and entered in the symbol library. This particular symbol corresponds to the memory cell shown in Fig.10.9

one or both of the display windows and, in fact, long lines may be drawn for
distances as great as desired using both SCROLL and DRAW simultaneously.

In addition to the major commands mentioned above, ICARUS has certain utility commands such as LIST for providing an inventory of predefined symbols, GET DRAWING for retrieving a design from disk, SAVE DRAWING for storing work in progress, and PRINT for generating a drawing of the design on either a Xerox laser printer or on a Versatec. The INSERT TEXT command permits alphanumeric data to be added to the drawing (without appearing on the final mask-generating tapes). Finally the QUIT command terminates a session. The resultant design may be checked by means of an off-line design rule checking program. Circuit simulation facilities are also available using SPICE. The final output of ICARUS is a file which can be converted to the tapes for mask generation and semiconductor fabrication.

10.7.4 Design Results

The resultant LSI/PHP chip is a four-bit version of the original sixteen-bit PHP. As described above, four PHP chips are to be interfaced to the Versabus of the Motorola 68000 in order to carry out cellular logic operations on sixteen picture elements simultaneously. The block diagram shown in Fig.10.11 indicates the major portions of the preliminary LSI/PHP design. Registers REG1, REG2, and REG3 are four-bit, on-chip memory registers which store data from three sequential lines of the binary image being processed. In addition, there are the B1, B2, and B3 one-bit memory registers needed to complete the neighborhood. Connections from these registers are made to the address lines of the four lookup tables (each of which requires a six-bit address). When the addressing cycle is complete and the outputs of the table are available, they are ANDed with the values of the center picture elements and the values of the mask register using the gating structure shown in Fig.10.12.

The actual chip layout in its preliminary form is shown in Fig.10.13. The total number of MOSFETs in this design is approximately four thousand so that in its current form it is an LSI design rather than a VLSI design. This indicates that further refinements employing the full power of VLSI may make it possible to generate a chip containing all sixteen lookup tables, their associated memory registers, and to some extent a portion of the image slice memory as well. Such a true VLSI design would have the obvious advantages of (1) decreasing the number of data transfers required over the Versabus while loading the memory registers; (2) placing the entire processor in a single chip; and (3) minimizing the total chip count, thereby increasing system reliability.

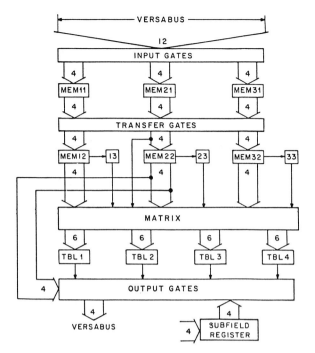

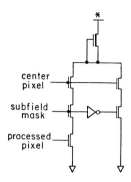

Fig.10.12. Information from the lookup tables, the center element in the hexagonal neighborhood, and the subfield mask is combined by the Kaufman gate shown schematically above

Fig.10.11. Block diagram of the LSI/PHP chip showing the memory registers, lookup tables, subfield register, and input/output gating. Data is passed 12 bits at a time from the Motorola 68000 over the Versabus to the memory registers. Offset addressing is carried out in the 68000. Lookup table addresses are generated on the chip and have six bits each. Lookup tables are loaded previously, one address at a time. Control is provided over three command lines whose contents are decoded to provide for the three major modes of operation

10.7.5 Operational Summary

The operation of the preliminary design of the LSI/PHP may be illustrated by numbering the picture elements in the 3 × 12 slice of image data as shown in Fig.10.14. Assume that initially the memory registers contain picture elements 1-2-3-4 in REG1, 13-14-15-16 in REG2, and 26-27-28-29 in REG3. Simultaneously, the Motorola 68000 would place input for elements 5-6-7-8 on inputs IMG1; 17-18-19-20 on IMG2; 30-31-32-33 on IMG3. At this point in the image processing cycle, output data for image elements 13-14-15-16 are calculable except that the entire neighborhood for image element 13 is unavailable due to its position on an image border. In the next step in the processing sequence, image elements 9-10-11-12 would appear at IMG1; 21-22-23-24 at IMG2; 34-35-36-37 at IMG3. At the same time, elements 5-6-7-8 would be stored in REG1; 17-18-19-20 in REG2; 30-31-32-33 in REG3. Finally, element 4 would be stored

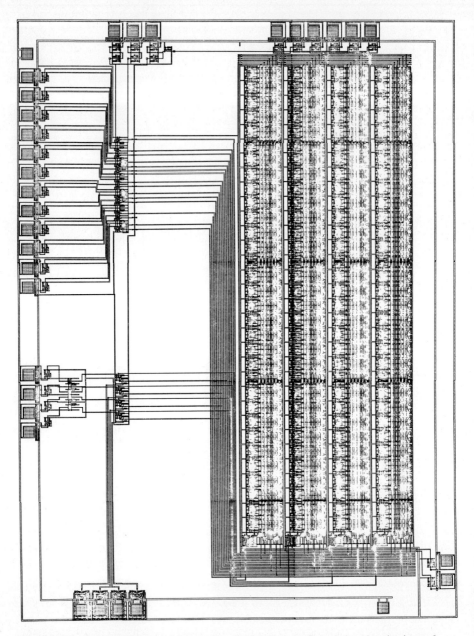

Fig.10.13. Overall design layout for the masks employed in the design of the preliminary LSI/PHP chip. Open areas are reserved for future components to complete the interface between the Motorola 68000 microprocessor and the input and output registers

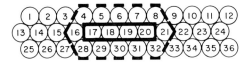

Fig.10.14. Picture element layout of a 3 × 12 section of image data to illustrate the flow of this data through the LSI/PHP cellular logic processor

in register B1; 16 in B2; and 29 in B3. At this time, all data necessary to compute the output values of elements at 17-18-19-20 would be available, namely, the elements 5-6-7-8 in REG1 and element 4 in B1; elements 17-18-19-20 in REG2 and element 16 in B2; elements 30-31-32-33 in REG2 and element 29 in B3; and, finally, element 21 at the IMG2 input. Thus, all four 6-bit addresses would be simultaneously available for manipulating the lookup tables. The outputs of the lookup tables would then be combined with the values of the central pixel elements (17-18-19-20) in the output and subfield masking gates to place the output data on the Versabus.

Acknowledgements. The authors would like to acknowledge the assistance of Ms. Rhonda Chandler, Ms. Angel Gordon, and Ms. Anne Volckmann, of Executive Suite, Inc., Tucson, Arizona, for typing this manuscript; of Mr. G. Thomas and his photographic design and drafting group at the Mellon Institute for doing the illustrations, and of Dr. J.M. Herron, Department of Radiation Health, University of Pittsburgh, for assisting with and critiquing the design. This research was supported in part by a grant (ECS81-13530) from the Division of Electrical, Computer, and Systems Engineering of the United States National Science.

References

10.1 J. von Neumann: The general logical theory of automata, in *Cerebral Mechanisms in Behavior - The Hixon Symposium*, ed. by L.A. Jeffress (Wiley, New York 1951)
10.2 M.J.B. Duff: Cellular logic array for image processing, Pattern Recognition **5**, 229-246 (1973)
10.3 K.E. Batcher: Design of a massively parallel processor, IEEE Trans. C-**29**, 836-840 (1980)
10.4 K. Preston, Jr.: "Machine Techniques for automatic identification of the binucleate lymphocyte", Proc. 4th Intern. Conf. Medical Electronics, Washington, DC (1961)
10.5 D. Graham, P.E. Norgren: The diff3 analyzer: A parallel/serial Golay image processor, in *Real-Time Medical Image Processing*, ed. by M. Onoe, K. Preston, Jr., and A. Rosenfeld (Plenum, New York 1980)
10.6 S.R. Sternberg: Parallel Architectures for Image Processing, in *Real-Time/Parallel Computers: Image Processing*, ed. by M. Onoe, K. Preston, Jr., and A. Rosenfeld (Plenum, New York 1981)
10.7 K. Preston, Jr.: Ξ-filters, IEEE Trans. ASSP (August 1983)
10.8 K. Preston, Jr.: Multidimensional logical transforms, IEEE Trans. PAMI (October 1983)
10.9 N.F. Izzo, W. Coles: Blood cell scanner identifies rare cells. Electronics **35**, 52-57 (April 1962)
10.10 M.J.E. Golay: Hexagonal parallel pattern transformations. IEEE Trans. C-**18**, 733-740 (1969)

10.11 K. Preston, Jr.: Application of cellular automata to biomedical image processing, in *Computer Techniques in Biomedicine and Medicine* (Auerbach Publishers, Philadelphia 1973)
10.12 J.M. Herron, J. Farley, K. Preston, Jr., H. Sellner: A general-purpose high-speed logical transform processor. IEEE Trans. C-**31**, 795-800 (1982)

11. Design of VLSI Based Multicomputer Architecture for Dynamic Scene Analysis

D.P. Agrawal and G.C. Pathak

With recent advances in VLSI technology it has now become feasible to design and implement computer systems incorporating a number of processors. The suitability of such systems for specific applications largely depends on some sort of match between the architecture and the algorithm(s) to be implemented. This work provides an objective approach of assessing the performance of an architecture for real-time dynamic scene analysis and outlines the design of an optimal VLSI-based system for this application.

The basic strategy employed partitions the serial algorithm for computer vision into noninteractive independent subtasks so that "pseudoparallelism" could be employed at the subtask level. Scheduling of subtasks to various computing resources in a multicomputer system has been considered and performance evaluation of such an allocation for various multicomputer architectures has been undertaken in terms of speedup, processor utilization, communication channel utilization, and cardinality. For real-time computer vision application a VLSI-based multicomputer architecture, which also optimizes the turnaround time and the throughput, has been identified.

11.1 Background

A multicomputer system (MCS) is defined as a multiple number of independent computers connected via a communication network that constitutes the structure of such a sytem. Various strategies for designing MCSs have been covered in the literature [11.1]. Besides allowing sharing of resources and improved reliability, the main objective behind developing a MCS is to achieve a higher system throughput and faster computation. With the advance in VLSI technology it has now become feasible to design and implement computer systems having a large number of processors. Various architectures [11.2-4] have been proposed for MCS and regularity and modularity of these architectures not only provides ease of VLSI implementation but also helps in understanding the behavior of the system.

A critical step in the design and development of a distributed computer system is in finding the suitable representation of software and computer system [11.5] and allocation (matching) of the former to the latter. Many of the theoretical attempts include probabilistic load balancing [11.6] or pipelining the computation [11.7] in the computer system. The problem of scheduling is further accentuated by absence of any analytical comprehensive tool which could predict and provide relative performance measures for these systems.

This work concentrates on suitability of various multicomputer architectures for a specific application, dynamic scene analysis. To achieve this, the dynamic scene analysis algorithm has been transformed into noninteractive independent subtasks so that pseudoparallelism could be employed at the subtask level [11.8]. The algorithm has been represented in terms of computation flow graphs [11.9] which could also be called a macrodata flow graph. This definition is appropriate because the computation flow graph follows the pattern of data flow graphs with the exception that its nodes are a multiset of instructions and the graph is acyclic. An $O(n^2)$ algorithm has been used for scheduling the scene analysis software to various architectures, where n is the number of computers in the system.

The performance of such scheduling has been evaluated in terms of speedup, computing resource utilization, communication channel utilization and cardinality [11.9,10]. The scheduling algorithm presented here uses local maximization techniques to obtain global near optimum results.

Section 11.2 describes the dynamic scene analysis algorithm and its transformation to a computation flow graph. Section 11.3 enumerates some existing nontrivial regular architectures. Section 11.4 explains the scheduling algorithm and its parameters. Section 11.5 evaluates such scheduling by simulating the computer system.

11.2 Dynamic Scene Analysis Algorithm

In this section we describe the algorithm for extracting images of moving objects, discussed in detail in [11.8]. Here, a brief description of the modules has been presented to familiarize the reader with the application. Figure 11.1 shows the various modules of the algorithm with the associated data flow.

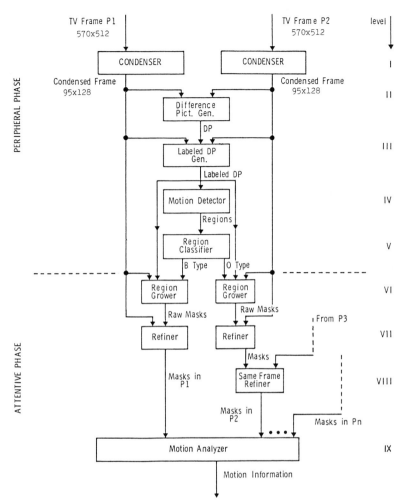

Fig.11.1. Sequential data flow and motion analyzer

11.2.1 Condensed Frame Generator

Input frames of 570 × 512 pixels are obtained using a television camera. A grey level of pixel is quantized to 256 levels. The input frames are condensed to a 95 × 128 picture by condensing four consecutive columns and five consecutive columns. Each element of the condensed frame contains the mean and variance of intensities of corresponding pixels of the original frame.

11.2.2 Difference-Picture Generator

A difference picture (DP) is a binary picture generated by comparing two continuous frames. For determining whether or not the corresponding pixels of the frames called previous and current frames may be considered different, R is computed as follows

$$R = [(S_p + S_c)/2 + (M_p + M_c)/2]^2 / S_p * S_c \quad , \tag{11.1}$$

where M and S denote mean and variance values contained at a pixel of condensed frame. Subscripts p and c denote previous and current frames. If R is greater than the threshold, then DP for the pixel will have the value 1, denoting the difference in the grey level; otherwise it will have the value 0.

11.2.3 Labeling

A labeled picture is obtained by applying the algorithm given below to every point (i,j) of DP, previous and current frames.

```
   begin
   if DP[i,j] = 1 then
     if E(DP[i,j]) then
        begin
        DP[i,j]: = 2;
        if SOB(PREV[i,j]) then DP[i,j]: = 3;
        if SOB(CURR[i,j]) then DP[i,j]: = DP[i,j] + 2;
        end;
end.
```

In this algorithm DP, PREV and CURR are the difference picture, previous frame and current frames and E and SOB are Boolean operators determining whether or not (i,j) is an edge point in binary and grey pictures, respectively.

11.2.4 Motion Detector

Difference pictures contain the changes in grey level in two continuous frames which could be caused either by motion of the object or noise. A filtering scheme eliminates the isolated changes which presumably are caused by noise. A connected component of more than say N elements is considered the result of motion.

11.2.5 Region Classifier

In this block three types of regions, formed due to occlusion or disocclusion or both, are determined and are named type O, B, and X regions, respectively. For a given region the type can be found by computing a ratio called CURPRE which equals the number as defined:

CURPRE: = # of points labeled 4 / # of points labeled 3.

The CURPRE is less than 1 for type B regions, greater than 1 for type O regions, and near 1 for type X regions.

11.2.6 Object Extractor

To obtain the masks in the present and the previous frames region growing is used. A region is grown by taking each nonregion pixel that has a horizontal neighbor within the region and comparing its gray level with that of an adjacent region pixel. If the gray levels are the same, then the nonregion pixel is added to the region. The gray levels are taken from the previous frame for the DP region of type B and from the current frame for type O. A similar process is applied for the nonregion pixels that have vertical neighbors within the region. These processes are iterated until no new pixels are added. The masks obtained in this way are further improved by using the refinement processes.

11.2.7 Motion Analyzer

The output of the object extractor are the images of the moving objects in continuous frames. Motion characteristics can be easily obtained from the displacement of the image.

11.2.8 Computation Flow Graph for Dynamic Scene Analysis

The computation flow graph (CFG) of an algorithm is a flow graph with source and sink nodes. Each node of the CFG represents a multiset of instructions which receive the data from the predecessor nodes and the resulting data is transferred to the follower nodes. A number written on the left of each node denotes the amount of computation at that node. A directed link between two nodes represents the data transfer and quantity of data is written on the link.

Figure 11.2 shows the CFG for a dynamic scene analysis algorithm. The complete frame is partitioned into four parts thus limiting the degree of parallelism to only four (it could be increased) for our purpose. Furthermore, an assumption has been made that in the whole picture we have only two regions

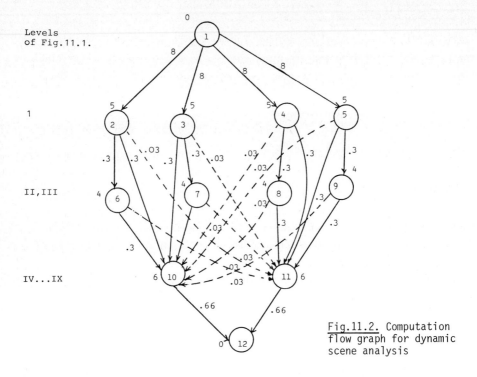

Fig.11.2. Computation flow graph for dynamic scene analysis

as shown by Nodes 10 and 11 and only 10% of the data is to be transferred from one region growing node to another region growing node. This overlapping data is shown by dotted lines in the figure. It is quite a realistic assumption and is rather overly estimated. As shown in the figure, Node 1 has no computation but transfers the data to Nodes 2, 3, 4 and 5. Node 6 has 4 units of computation and transfers data to Nodes 10 and 11. In the present graph one unit of computation corresponds to one hundred thousand instructions.

11.3 Existing Multicomputer Architecture

There are many interconnection methods, or topologies, for linking the network of computers [11.2-4]. In this contribution, we avoid the trivial networks like star, circular ring and binary tree because they weaken the concept of distributed computing by having a low fault tolerance. Various candidate architectures for dynamic scene analysis applications have been discussed in the following paragraphs. *Wittie* has presented a detailed description of the relative merits of these networks in [11.2].

11.3.1 Full Connected Network

In a full connected network every pair of nodes is provided with a dedicated link. These networks are rarely used since line and connection costs grow as n^2. Figure 11.3a shows a full connected network of twelve nodes.

11.3.2 Cube Connected Cycles

The cube connected cycle (CCC), a hypercube topology, is suitable for very large networks. This has been shown to be very time efficient for a large number of distributed algorithms and to have an area efficient for VLSI layout [11.11].

A CCC of dimension D consists of $D * 2^D$ nodes arranged as a cycle of D nodes around each of 2^D vertices of a binary hypercube of D dimensions. Each node is connected to exactly three others by dedicated bidirectional links. Figure 11.3b shows a CCC network of twenty four nodes.

11.3.3 Alpha Network

The alpha structure proposed by *Bhuyan* and *Agrawal* [11.12] can be viewed as a hypercube structure of r dimensions with m_i nodes in the ith direction, where a node in a particular axis is connected to all other nodes in the same axis.

The total number of nodes n is represented as a product of m_is for $1 < i < r$

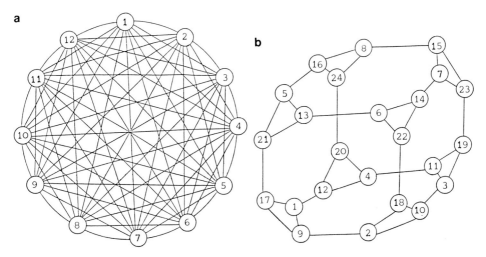

Fig.11.3a. Fully connected network. (b) Cube connected cycle network

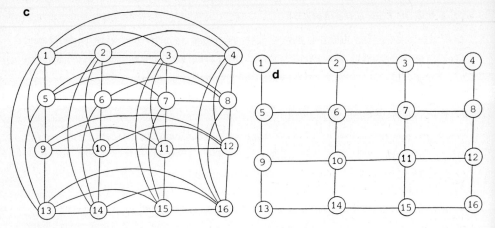

Fig.11.3c. Alpha structure. (d) Mesh connected network

$$n = m_1 \times m_2 \times \ldots \times m_r \quad .$$

Each node is connected to the nodes having hamming distance of one in radix r. The alpha network of sixteen nodes is shown in Fig.11.3c.

11.3.4 Mesh Connected Network

Illiac IV, a widely known SIMD machine, employs a mash connection scheme for the slave processing elements. The regular nearest-neighbor array or W-wide D-dimensional mesh consists of $n = W^D$ nodes. Figure 11.3d shows a two-dimensional mesh with sixteen nodes. In this chapter we have not used the end-around connection, in order to maintain an expandable structure at the cost of increasing the average distance.

11.3.5 Hypertree Topology

Goodman and *Sequin* [11.4] proposed a hypertree topology which combines the features of n-cube connections with a binary tree, thereby improving the average distance and fault tolerance of the system. Hypertree is a binary tree with additional horizontal links connecting the nodes of the same level. In particular, they are chosen to be a set of n-cube connections, with only one part per node available (in our case, the choice being made to reduce the longest connection at each level. Figure 11.3e shows a hypertree connection network consisting of fifteen nodes.

e

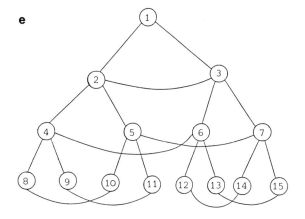

Fig.11.3e. Hypertree structure

11.3.6 Multitree Structure

Proposed by *Arden* and *Lee* [11.3], a multitree structure (MTS) of size (m,t) is defined as follows

there are m identical CTs of depth (t-1);
the root of m CTs are interconnected to form a ring;
for level (t-1) each node is connected to (d-1) other level (t-1) nodes and there is at least one cycle containing all the level (t-1) nodes.

f

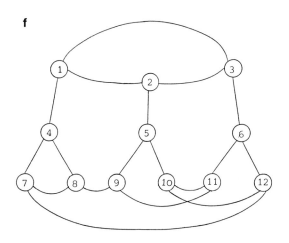

Fig.11.3f. Multitree structure

The component tree (CT) of an MTS graph of degree d is a rooted undirected tree where each nonleaf node has (d-1) sons except the root which has (d-2) sons.

Figure 11.3f shows an MTS connection topology for twelve nodes.

11.3.7 Computing Resource Graph

In this chapter, a multicomputer system is represented by a computing resource graph (CRG). A CRG of a MCS is an undirected graph in which each node denotes a computer and a link between two nodes shows that a communication link exists between these two nodes. A number written on the left of the node is the computing power (usually instructions/second) of each computer. In general, a number of attributes could be attached to each node, e.g., memory. The number written on the links represents the physical link bandwidth.

Figure 11.4 shows a CRG graph for the hypertree structure. The number on the top left of the node denotes the computing power of the node. In the figure the computing power unit is a thousand instructions per second. The number written on the link denotes the data transfer capability of the link. Node 1 can compute 100 instructions per second and transfers the data to Nodes 2 and 3 at the rate of 50 K bytes/second.

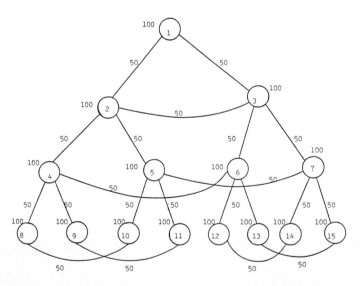

Fig.11.4. Computing resource graph for hypertree computer system

11.4 Scheduling and Parameters of Interest

In CFG of dynamic scene analysis, as shown in Fig.11.2, computation at each node follows a precedence rule. If there is a directed edge from node i to node j then node j can be computed only after node i has been computed and data transfer from node i to node j has taken place. On the basis of this relationship we define the term level of each node i as

level i = 1 if i is the root node else
       = V max (level[pred(i)]) + 1;

where pred(i) is the predecessor of the node i.

This implies that the computation nodes of the same level are independent and so could be computed concurrently.

In the present work we have used a scheduling algorithm, which utilizes the local search method, for matching the CFG to various architectures. Although an exhaustive search algorithm would provide optimum results, due to exponential time complexity it becomes impractical to use in real time for large networks. The solutions presented in this paper cannot be taken as optimal but they conform to the theoretical expectations, as will be shown in Sect.11.5.

The algorithm selects an unmatched CFG node from the current level, depending upon which node has a maximum amount of computation and data communication. A match for this node is searched in the CRG of the computer system under investigation. The matching which minimizes the optimization function is marked and this CRG node is deleted from the CRG graph. One must not forget that CRG is a fully connected graph on the logical links. So the remaining graph is still connected. The optimizing function has few constants whose values can be changed to emphasize the communication or computation in the CFG to suit the scheduling algorithm for communication—or computation-intensive applications.

Various parameters of interest when matching are:

1) turnaround time,
2) speedup,
3) computer utilization,
4) channel utilization,
5) cardinality.

Turnaround time reflects the time taken by the system to process the ith frame of the scene analysis algorithm. Speedup shows how quickly the next frame can be fed for processing. It is a relative measure and the reference is a unicomputer system. Computer utilization represents the amount of time computers are used for computation: the higher the computer utilization, the better the system. Similarly channel utilization represents the effective use of the communication channels. Cardinality of the matching shows the number of CFG edges that fall on the physical links of the system in the scheduling. In the next section we analyze the performance of various computer architectures for image scene analysis.

11.5 Performance Evaluation

We have implemented a program which simulates the multicomputer system and produces the required parameters. The input to this program is the computation flow graph, computing resource graph and the mapping. Incidence matrices are generated for a given network and for a given number of nodes. These incidence matrices are fed to a program which produces the logical link structures. The assumption has been made that the shortest distance between two nodes determines the data communication bandwidth of these nodes. A scheduling algorithm matches the CRG thus generated to the CFG of scene analysis problems. Our simulation program uses the event simulation method to represent the multicomputer system.

Table 11.1 shows the result of scheduling various nontrivial architectures for dynamic scene analysis. As expected, the cardinality of the fully connected system exceeds any other (shown in the first column): CCC has the least cardinality. Cardinality does not prove any structure superior or inferior. It shows rather the number of direct (without routing) communications performed in the system. Turnaround time is obviously a dominant factor in evaluating a network. Fully connected and alpha structures provide better turnaround time. Speedup turns out to be the same for all the cases because of the way it has been defined in the simulation program. The speedup is the maximum computation time in all the levels, also termed "pipeline speed" in the literature. It has been assumed that communication overhead is dealt with separately which is not always true. Computer and channel utilization denote the use of resources. Higher values of these factors imply better system performance, while degree of network interconnection implies the communication cost, where fully connected networks are worst in this aspect. The last column of the table is the number of nodes we have taken for a particular network. Depending upon the weightage scheme (the term used by *Domenico* and *Jacob* [11.13], one can choose the system one intends to use.

Table 11.1. Performance of various architectures for dynamic scene analysis

Computer structures	Cardinality	Turnar. time	Speed-up	Comput. utzn.	Channel utzn.	Degree	Nodes
Alpha	13	0.34	80	1.4	2.2	6	16
Hypertree	10	0.53	80	0.91	2.2	4	15
Full	26	0.34	80	1.4	2.2	11	12
CCC	6	0.53	80	0.91	1.9	3	24
Mesh	8	0.51	80	0.91	2.2	4	16
MTS	13	0.51	80	0.94	1.9	3	12

11.6 Conclusion

In this chapter we have outlined a method of comparing the performance of structure for the dynamic scene analysis problem. The dynamic scene analysis algorithm has been transformed to a CFG which captures the essentials of the algorithm. We experimented on a limited scale to analyze the results. A detailed scheduling can be designed on similar lines to include special-purpose computers and memory constraints. Suitability of a structure to VLSI is judged by its regular and extendable structure. We have taken care to include such structures as possible candidates for dynamic scene analysis.

References

11.1 M.O. Ward: The automated design of task specific parallel processing architectures, Proc. Int. Conf. on Parallel Processing (1982) pp. 298-300
11.2 L.D. Wittie: Communication structure for large network of microprocessors, IEEE Trans. C-**30**, 264-273 (1981)
11.3 B.W. Arden, H. Lee: A regular network for multicomputer system, IEEE Trans. C-**31**, 60-69 (1982)
11.4 J.R. Goodman, C.H. Sequin: Hypertree: A multiprocessor interconnection topology, IEEE Trans. C-**30**, 923-933 (1981)
11.5 M.J. Flynn, J.L. Hennessy: Parallelism and representation problem in distributed systems, Conf. on Dist. Comput. Systems, Huntsville, AL (1979) pp.124-130
11.6 L.M. Ni, K. Hwang: Optimal load balancing strategies for a multiprocessor system, Proc. Intern. Conf. Parallel Processing, (1981) pp.352-357
11.7 D. Dours, R. Facca: Multiprocessor architecture adapted to the parallel treatment of a continuous flow of data, 5th Intern. Conf. on Pattern Recognition, Miami, FL (1980) pp.321-325
11.8 R. Jain, W. Martin, J.K. Agrawal: Segmentation through the detection of change due to motion, Comput. Graphics Image processing **11**, 13-34 (1979)

11.9 D.P. Agrawal, G.C. Pathak: Performance of multicomputer architectures, SIAM Special Conf. on Parallel Processing, Norfolk (1983)
11.10 S.H. Bokhari: A shortest tree algorithm for optimal assignment across space and time in a distributed processor system, IEEE Trans. SE-7, pp.583-589 (1981)
11.11 F.P. Preparata, J. Vuillemin: The cube connected cycle: A versatile network for parallel computation, in Proc. 20th Symp. on Found. of Comput. Sci. (1979) pp.140-147
11.12 L.N. Bhuyan, D.P. Agrawal: A general class of processor interconnection strategies, Intern. Symp. on Computer Arch. (1982) pp.90-98
11.13 D.M.D. Domenico, S.M. Jacob: A method for comparing distributed computer system architecture, Proc. of Computer Software and Applications Conf. (1982) pp.78-83

12. VLSI-Based Image Resampling for Electronic Publishing

Z.Z. Stroll and S.-C. Kang

12.1 Introduction to the "Electronic Darkroom"

With the advent of the invasion of the office and publishing by electronics, new product opportunities have opened, and many have begun to be exploited. These include word, text, and graphics processing among a myriad of other applications. Whether locally or remotely generated or updated, data is stored, mailed, distributed, printed, edited, etc. Conspicuously absent from this battleground has been the handling of image data in a graphics composition/page makeup environment due to the prohibitive costs involved with scanning, storing, processing, and printing of scanned imagery. But this last bastion is now under attack and will fall in the near future to the onslaught of advances in very large scale integrated electronics, novel storage and maturing scanning and marking technologies.

A set of functions exist that are better handled by operating on the electronic master [12.1], including (but not limited to) reduction, magnification, rotation, filtering, parametric and nonparametric cropping, halftoning/TRC modification, etc. The functions above can be selectively applied to the electronic master without obliterating it and the results interactively evaluated for adequacy. The first four can be accomplished by resampling the electronic master to the resized, rotated coordinate system using a loadable resampling function kernel.

The impetus to image resampling was provided during the last decade by NASA needs in removing satellite and scanning system errors from the LANDSAT imagery beamed back to earth [12.2,3]. Procedures were developed to estimate the error distribution profile followed by an image resampling process to warp the image back into the correct shape in a variety of coordinate systems. LANDSAT image data resampling requires a high degree of spatial and radiometric accuracy, a small amount of rotation (skew of up to $1.5°$) and a reasonable processing rate (2 to 10 μs/resampled pixel).

Theoretically, convolution with the sinc function provides ideal resampling [12.4]; unfortunately this requires processing a very large number of pixels. A truncated sinc function has been accepted as the quality resampling standard, with bilinear and nearest neighbor as standby options. The resampling process involves four nearest-neighboring pixels (two on each side) in each direction, hence a single-pass interpolation requires 16 multiply and add operations per resampled pixel, while a two-pass system requires 4 multiply and add operations in each pass. Implementations are based on special purpose or specially programmed floating/fixed point signal processors.

The introduction of resampling in commercial applications has been lagging behind defense/space requirements; however, demand is now growing due to the increasing sophistication and technological advances that are being achieved in advanced systems products targeted at the office, communications and publishing industries. The commercial image data resampling requirements will have somewhat different characteristics than the ones formerly described. Sufficient quality can be obtained by using a 2-point (4-point two-dimensional) resampling process with a loadable kernel function to provide certain filtering capabilities. However, the dynamic range requirements for rotation and scaling are much greater, necessitating the judicious use of system resources such as mass storage, main processor data buffering, private data storage and arithmetic operators. While pixel processing rates at the hardware level can be feasibly accomplished at 0.5 to 1 μs/pixel, mass storage access may restrict the overall rate by a factor from four to eight. Though the resampling process could also accommodate the removal of scanning system nonlinearities, it is anticipated that these functions will be performed in situ at the scanner.

Image cropping is accomplished through the dual functions of parametric cropping of the edges, along with the superposition of a composition bitmap for nonparametric cropping. The composition bitmap is at the CRT display resolution and size, and arbitrary resolution ratios among scanner, display and marker must be handled. Scanned image decompression and decoding as well as halftoning for black/white display or marking are also accommodated.

12.2 Overview of System Design Concepts

This study investigates two approaches to image resampling: a single-pass and a two-pass method. At the heart of the single-pass resampler is the four-point two-dimensional interpolator.

The original data scan lines are retrieved from disk, buffered in main memory, then segments along the new scan direction are shipped to the resampling hardware. Here the four-point interpolator resamples the data into com-

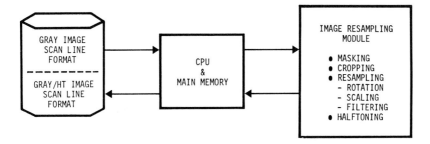

Fig.12.1. Single-pass resampling functional data flow. *Note*: Along line image data decoding can be performed in line without rotation

plete new scan lines, subsequently shipping them back to the memory for buffering, then storage on disk.

The original scanned data on disk is assumed to be in gray format. The single-pass resampler can accommodate decoding of along-scan encoded data in the same operation only if no rotation is performed. This restriction does not pose much difficulty since the first operation would usually consist of decoding and conversion to CRT resolution in order to initiate interactive image manipulation. The functional data flow is shown in Fig.12.1. Thereafter, much of the interactive operation would be performed on the CRT resolution display master. The former is not a requirement, though processing the display master is considerably faster than always returning to the full resolution gray master. For finished copy generation, of course, all of the processing parameters are applied to the full resolution master.

The single-pass resampler, in addition to scaling, rotation and filtering, accommodates all the "electronic darkroom" functions such as the application of the composition bitmap, parametric cropping and halftoning. The composition bitmap is aligned with the image before rotation. The mask is superimposed on the data before resampling. Cropping parameters are specified in terms of the resampled grid units and directions. The halftone cells are likewise aligned with the resampled grid.

The two-pass resampler is so named because the image data passes twice through the resampling hardware, once for along line resampling, the second time for across line resampling, with intermediate storage on disk. The original data is retrieved from disk, buffered through main memory and shipped to the resampler hardware where along line resampling is performed by interpolating between two adjacent pixels on the scan line. Along line cropping and masking are also performed at this time. Subsequently, the data is returned in along line segments, each segment internally transposed to the

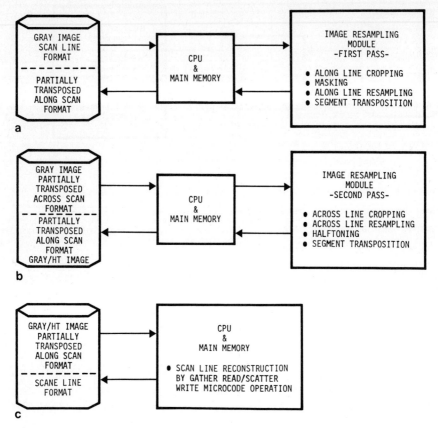

Fig.12.2a-c. Two-pass resampling functional data flow. a) First pass resampling operation; b) second pass resampling operation; c) optional third-pass scan line reformatting operation. *Note*: Along line image data can be decoded in line

across line direction. The buffered data is stored on disk in a format that allows efficient access to it in the across line segment order for the second pass. In the second pass, the segments are collected in across line sequence, and passed to the resampler for across line resampling in an analogous fashion to the along line process described above. Across line cropping and halftoning are also performed at this time. Then the segments, optionally transposed internally for rotation angles of $45°$ or less, are returned to be buffered through main memory and stored on disk. Unlike the one-pass resampler, the two-pass resampler does not restrict rotation while decoding. The functional data flow diagram is given in Fig.12.2.

12.3 Resampling Algorithms

The image resampling process consists of a coordinate transformation that determines the location of the resampled pixel in the original image space and an interpolation of neighboring pixels to obtain the resampled pixel value at the desired location. The coordinate transformation which is involved in rotation and scaling can be described in matrix notation as follows: let the original coordinates be x,y and the resampled coordinates be X,Y. Then

$$[X\ Y\ 1] = [x\ y\ 1]M$$

where the coordinate transformation matrix,

$$M = M_1 \ldots M_i \ldots M_k = \begin{bmatrix} a & d & 0 \\ b & e & 0 \\ c & f & 1 \end{bmatrix}$$

is the concatenation of a series of elementary operators such as

$$M_i = \begin{bmatrix} C & S & 0 \\ -S & C & 0 \\ 0 & 0 & 1 \end{bmatrix}, \quad C = \cos\theta, \quad S = \sin\theta$$

for rotation transformation, or

$$M_i = \begin{bmatrix} S_x & 0 & 0 \\ 0 & S_y & 0 \\ 0 & 0 & 1 \end{bmatrix}$$

for scaling transformation, or

$$M_i = \begin{bmatrix} 1 & 0 & 0 \\ 0 & 1 & 0 \\ T_x & T_y & 1 \end{bmatrix}$$

for translation transformation, etc. Thus,

$$[x\ y\ 1] = [X\ Y\ 1]M^{-1}$$

$$= [X\ Y\ 1] \begin{bmatrix} \Delta X_x & \Delta Y_x & 0 \\ \Delta X_y & \Delta Y_y & 0 \\ SP_x & SP_y & 1 \end{bmatrix}, \qquad (12.1)$$

where

$\Delta X_x = e/D$ $\qquad\qquad \Delta Y_x = -d/D$

$\Delta X_y = -b/D$ $\qquad\qquad \Delta Y_y = a/D$

$SP_x = (bf - ce)/D$, $\quad SP_y = (cd - af)/D$, $\quad D = ae - bd$.

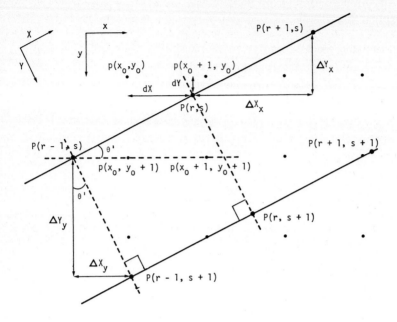

Fig.12.3. Geometry of single-pass resampling

The geometrical significance of these parameters can be seen in Fig.12.3.

Let an array of $m \times n$ pixels in the $[x\ y]$ coordinate system, with pixel values at (i,j) denoted by $p(i,j)$, be defined for the integers $0 \leq i \leq n-1$, $0 \leq j \leq m-1$ to represent the original image. The pixel value $p(x,y)$ at (x,y) is obtained by convolution with the two-dimensional interpolation kernel function $W(x,y)$

$$p(x,y) = \sum_{i=0}^{n-1} \sum_{j=0}^{m-1} p(i,j) W(x-i, y-j) \quad . \tag{12.2}$$

The above formula is simplified using an orthogonally separable interpolation function for which $W(x,y) = W(x)W(y)$ and interpolating only l neighbors. That is,

$$p(x,y) = \sum_{i=0}^{l-1} \sum_{j=0}^{l-1} p(x_i, y_j) W(x - x_i) W(y - y_j) \quad , \tag{12.3}$$

where x_i and y_j are one of the l nearest integers of x and y, respectively; $p(x,y)$ has a "background" value for the pixels undefined in the given pixel array. Then the resampled pixel at the grid point (r,s) in the $[X\ Y]$ coordinate system is $P(r,s) = p(x_r, y_s)$ where $[x_r\ y_s\ 1] = [r\ s\ 1]M^{-1}$, $x_r = x_0 + dX$, $y_s = y_0 + dY$, $l = 2$ and

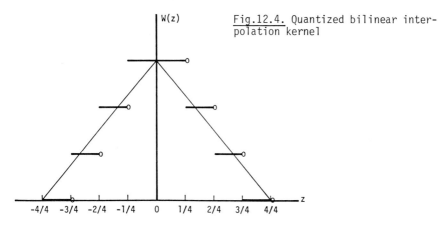

Fig.12.4. Quantized bilinear interpolation kernel

$$p(x_r, y_s) = p(x_0, y_0)W(dX)W(dY)$$
$$+ p(x_0 + 1, y_0)W(dX - 1)W(dY)$$
$$+ p(x_0, y_0 + 1)W(dx)W(dY - 1)$$
$$+ p(x_0 + 1, y_0 + 1)W(dX - 1)W(dY - 1) \quad . \tag{12.4}$$

The above forms the basis of a one-pass resampler. The mechanization involves only registers and adders to allow indexing through the original pixel space along the resampled scan line. Moreover, judicious rounding of parameters in M^{-1} and the use of limited significance integer values yields a system capable of meeting required spatial accuracy. When the interpolation function contains a limited number of discrete values, multiplication can practically be implemented by table lookup using for instance a quantized bilinear interpolation kernel function as shown in Fig.12.4. Since the pixel space is closed, the multiplication can be iterated via the same lookup process.

An alternate approach to performing the coordinate transformation forms the foundation of the two-pass resampler as follows: let

$$[X \; Y \; 1] = [x \; y \; 1]M = [x \; y \; 1]M_x M_y$$

$$= [x \; y \; 1]\begin{bmatrix} a & 0 & 0 \\ b & 1 & 0 \\ c & 0 & 1 \end{bmatrix}\begin{bmatrix} 1 & d' & 0 \\ 0 & e' & 0 \\ 0 & f' & 1 \end{bmatrix} \quad , \quad \begin{array}{l} d' = d/a, \; a \neq 0 \\ e' = e - bd' \\ f' = f - c'd' \end{array} \quad .$$

Then

$$[x \; y \; 1] = [X \; Y \; 1]M_y^{-1}M_x^{-1}$$

215

$$= [X\ Y\ 1] \begin{bmatrix} 1 & \Delta Y_2 & 0 \\ 0 & \Delta X_2 & 0 \\ P & SP_2 & 1 \end{bmatrix} \begin{bmatrix} \Delta X_1 & 0 & 0 \\ \Delta Y_1 & 1 & 0 \\ SP_1 & 0 & 1 \end{bmatrix}$$

$$= [X\ Y\ 1] PM_y^{-1} PM_x^{-1} \quad ,$$

where

$$M_{y'}^{-1} = PM_y^{-1}P, \quad P = \begin{bmatrix} 0 & 1 & 0 \\ 1 & 0 & 0 \\ 0 & 0 & 1 \end{bmatrix} \quad .$$

Consequently,

$$[x\ y\ 1] = [X\ Y\ 1] \begin{bmatrix} 0 & 1 & 0 \\ 1 & 0 & 0 \\ 0 & 0 & 1 \end{bmatrix} \begin{bmatrix} \Delta X_2 & 0 & 0 \\ \Delta Y_2 & 1 & 0 \\ SP_2 & 0 & 1 \end{bmatrix} \begin{bmatrix} 0 & 1 & 0 \\ 1 & 0 & 0 \\ 0 & 0 & 1 \end{bmatrix} \begin{bmatrix} \Delta X_1 & 0 & 0 \\ \Delta Y_1 & 1 & 0 \\ SP_1 & 0 & 1 \end{bmatrix} \quad ,$$

where

$\Delta X_1 = 1/a \qquad \Delta X_2 = 1/e'$

$\Delta Y_1 = -b/a \qquad \Delta Y_2 = -d'/e'$

$SP_1 = -c/a \qquad SP_2 = -f'/e' \quad .$

The geometric significance of the parameters above can be seen in Fig. 12.5. The two-pass resampling process pictured in Fig.12.6 can be described as follows. In the first pass, an intermediate image is determined in the

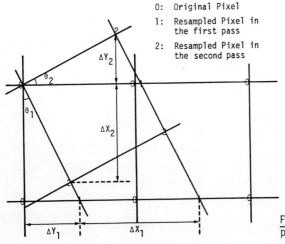

Fig.12.5. Geometry of two-pass resampling

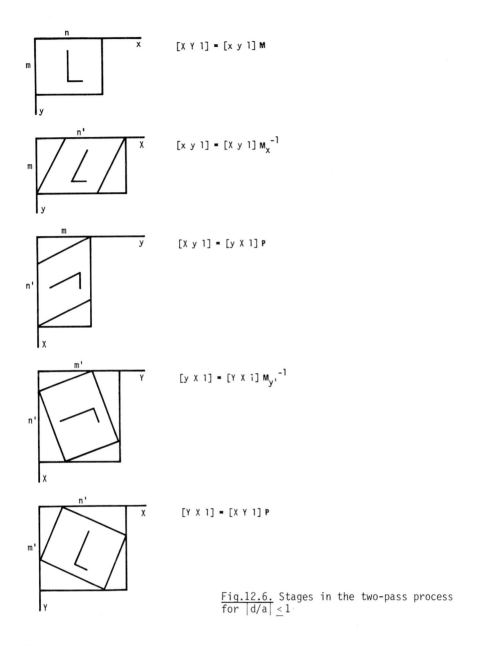

Fig.12.6. Stages in the two-pass process for $|d/a| \leq 1$

[X y] coordinate system as a rectangular array of $m \times n'$ pixels. The (x,y) coordinates for the new pixels defined in the [X y] coordinate system are obtained using $[x\ y\ 1] = [X\ y\ 1]M_x^{-1}$. Note that the first pass resampling parameters, ΔX_1, ΔY_1 and SP_1, are defined by M_x^{-1}. Each pixel value at (X,y) is obtained interpolating original pixels around (x,y). The intermediate image is then transposed in order to resample in the across scan line direction afterwards. This process is represented in terms of a multiplication by a permutation matrix **P**, i.e., [X y 1] = [y X 1]**P**. The second pass can be accomplished similarly. The final image is defined in the [X Y] coordinate system as an array of m' × n' pixels. The (y,X) coordinates for the pixels defined in the [X Y] coordinate system are obtained using [y X 1] = [Y X 1]$M_{y'}^{-1}$ and [Y X 1] = [X Y 1]**P**. The final pixel values at (X,Y) are obtained interpolating pixels around (X,y). The second pass resampling parameters, ΔX_2, ΔY_2, SP_2, are defined by $M_{y'}^{-1}$.

One point should be noted in two-pass resampling. When the component "a" of the matrix **M** has a small magnitude, the resampling parameters, $\Delta X_1 (=1/a)$ and $\Delta Y_1 (=-b/a)$ have large magnitudes. This would cause the pixels in the final image to be obtained interpolating original pixels which are not close to the resampling positions. This effect can be alleviated if the transformation matrix is permuted before resampling when the component "d" of the matrix **M** has a larger magnitude than the component "a". Thus, when $|d/a| > 1$, the resampling parameters are derived as follows:

[X Y 1] = [x y 1]**M**

= [x y 1]**M P P**, **P P** = **I** = Identity matrix .

$$[X\ Y\ 1] = [x\ y\ 1] \begin{bmatrix} a & d & 0 \\ b & e & 0 \\ c & f & 1 \end{bmatrix} \begin{bmatrix} 0 & 1 & 0 \\ 1 & 0 & 0 \\ 0 & 0 & 1 \end{bmatrix} \begin{bmatrix} 0 & 1 & 0 \\ 1 & 0 & 0 \\ 0 & 0 & 1 \end{bmatrix}$$

$$= [x\ y\ 1] \begin{bmatrix} d & a & 0 \\ e & b & 0 \\ f & c & 1 \end{bmatrix} \begin{bmatrix} 0 & 1 & 0 \\ 1 & 0 & 0 \\ 0 & 0 & 1 \end{bmatrix}$$

$$= [x\ y\ 1] N_x N_y P$$

$$= [x\ y\ 1] \begin{bmatrix} d & 0 & 0 \\ e & 1 & 0 \\ f & 0 & 1 \end{bmatrix} \begin{bmatrix} 1 & a' & 0 \\ 0 & b' & 0 \\ 0 & c' & 1 \end{bmatrix} \begin{bmatrix} 0 & 1 & 0 \\ 1 & 0 & 0 \\ 0 & 0 & 1 \end{bmatrix} , \quad \begin{aligned} a' &= a/d,\ d \neq 0 \\ b' &= b - ea' \\ c' &= c - fa' \end{aligned} .$$

The old coordinate system [x y] can be obtained from [X Y] as follows:

$$[x\ y\ 1] = [X\ Y\ 1] P\ N_y^{-1}\ N_x^{-1},\ P^{-1} = P\ .$$

Consequently,

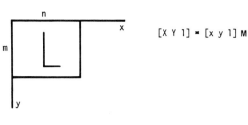
$$[X\ Y\ 1] = [x\ y\ 1]\ \mathbf{M}$$

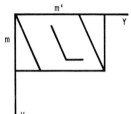
$$[x\ y\ 1] = [Y\ y\ 1]\ \mathbf{N}_x^{-1}$$

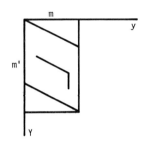
$$[Y\ y\ 1] = [y\ Y\ 1]\ \mathbf{P}$$

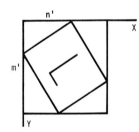
$$[y\ Y\ 1] = [X\ Y\ 1]\ \mathbf{N}_{y'}^{-1}$$

Fig.12.7. Stages in the two-pass process for $|d/a| > 1$

$$[x\ y\ 1] = [X\ Y\ 1] \begin{bmatrix} 0 & 1 & 0 \\ 1 & 0 & 0 \\ 0 & 0 & 1 \end{bmatrix} \begin{bmatrix} 1 & \Delta Y_2 & 0 \\ 0 & \Delta X_2 & 0 \\ 0 & SP_2 & 1 \end{bmatrix} \begin{bmatrix} \Delta X_1 & 0 & 0 \\ \Delta Y_1 & 1 & 0 \\ SP_1 & 0 & 1 \end{bmatrix}$$

$$= [X\ Y\ 1]\mathbf{P}(\mathbf{PN}_{y'}^{-1}\mathbf{P})\mathbf{N}_x^{-1}$$

$$= [X\ Y\ 1]\mathbf{N}_{y'}^{-1}\mathbf{PN}_x^{-1}$$

$$= [X \ Y \ 1] \begin{bmatrix} \Delta X_2 & 0 & 0 \\ \Delta Y_2 & 1 & 0 \\ SP_2 & 0 & 1 \end{bmatrix} \begin{bmatrix} 0 & 1 & 0 \\ 1 & 0 & 0 \\ 0 & 0 & 1 \end{bmatrix} \begin{bmatrix} \Delta X_1 & 0 & 0 \\ \Delta Y_1 & 1 & 0 \\ SP_1 & 0 & 1 \end{bmatrix} ,$$

where

$\Delta X_1 = 1/d$ $\Delta X_2 = 1/b'$

$\Delta Y_1 = -e/d$ $\Delta Y_2 = -a'/b'$

$SP_1 = -f/d$ $SP_2 = -c'/b'$.

In the above case, the intermediate image as depicted in Fig.12.7 is defined in the [Y y] coordinate system as a rectangular array of $m' \times m$ pixels. The coordinates are obtained using $[x \ y \ 1] = [Y \ y \ 1]N_x^{-1}$. This image is then processed by using $[y \ Y \ 1] = [X \ Y \ 1]N_y^{-1}$. It is not required to transpose the image after second pass resampling since the obtained image is already in scan-line order. Mechanization aspects mentioned in the one-pass operation remain valid for the two-pass resampler.

12.4 System Architecture and Performance

The architectural block diagram of the image resampling system is shown in Fig.12.8. The resampling subsystem is embedded in the classical computer system model. The programmable microprocessor performs the overall process control, particularly as related to setting up disk, DMA and resampler data flows.

Let the resampled image be composed of a number of small blocks, Nblk. The following describes the one-pass process which finds the resampled image by obtaining Nblk's in sequence:

0) Read first L scan lines of the original image into the input buffer of main memory: L is determined by the size of the input buffer available.
1) Identify next Nblk to be obtained in the resampled scan direction (Fig. 12.9).
2) Find the original block of data, Oblk, containing the Nblk identified in Step (1). If the Oblk is not available in the input buffer of main memory, then read in another L scan lines of the original image.
3) Obtain Nblk by resampling Oblk. Store Nblk in the output buffer.
4) Repeat (1-3) until Nblk size (height of Nblk) scan lines of the resampled image are stored in the output buffer of main memory.

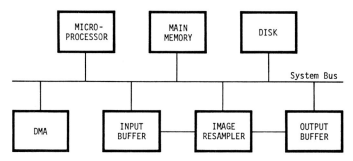

Fig.12.8. Resampling system block diagram

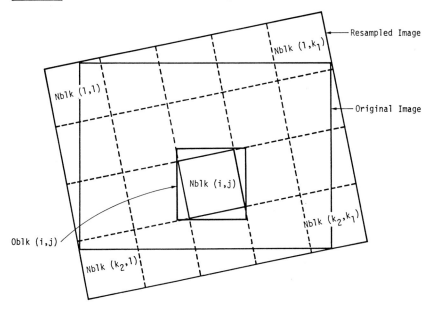

Fig.12.9. Resampled image blocks

5) Write output buffer to disk file.
6) Repeat (1-5) until the image is complete.

The two-pass process flow is as follows:

First Pass

1.1) Read H scan lines of the original image into the input buffer: H is determined by the buffer size available in main memory.
1.2) Resample H scan lines and transpose blocks (of size H × H).
1.3) Write transposed blocks to the intermediate disk file.
1.4) Repeat (1.1-3) until the image is complete.

Second Pass

2.1) Read H scan lines of the intermediate image.
2.2) Resample H scan lines and transpose blocks if required.
2.3) Write transposed blocks to the disk.
2.4) Repeat (2.1-3) until the image is complete.

If transposition is performed in the second pass, then the stored image is in a multiplexed scan-line format. The following steps can be used to demultiplex the scan lines of the final image:

3.1) Read H scan lines from the image obtained at the end of the second pass.
3.2) Demultiplex scan lines.
3.3) Write in the output file.
3.4) Repeat (3.1-3) until the image is complete.

The resampler subsystem functional flow diagram is shown in Fig.12.10 and that of the chip is shown in Fig.12.11. Figure 12.12 is a Pascal-like definition of the chip operation in both single and two-pass contexts.

The parameters involved in the position generation process are limited significance values represented by a fixed number of bits and the chip performs fixed point arithmetic on them. The roundoff errors involved in the quantities representing the increments in the next pixel position computation affect the two resampling algorithms in different manners. Figures 12.3,5 depict the differences in the resampling positions for a transformation involving rotation ($0 \leq \theta \leq 45°$) and scaling ($S_x = S_y = S$) only, i.e.,

$$M = \begin{bmatrix} S\cos\theta & S\sin\theta & 0 \\ -S\sin\theta & S\cos\theta & 0 \\ 0 & 0 & 1 \end{bmatrix} .$$

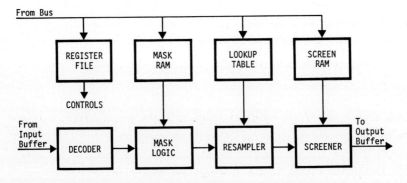

Fig.12.10. Resampler subsystem functional flow diagram

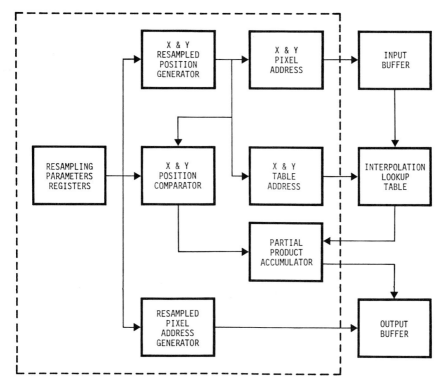

Fig.12.11. Resampler chip functional flow diagram

For the single-pass algorithm, $\Delta X_x = \Delta Y_y = \cos\theta'/S$ is obtained rounding off the value $\cos\theta/S$ and $\Delta X_y = -\Delta Y_x = \sin\theta'/S$ from $\sin\theta/S$ where $\theta' = \tan^{-1}(\Delta X_y/\Delta X_x)$ is the actual angle of rotation instead of θ and $S' = (\Delta X_x^2 + \Delta Y_x^2)^{-\frac{1}{2}}$ is the actual scale factor obtained instead of S. Note that the resampled grid forms a square grid.

For the two-pass algorithm, four increments, ΔX_1, ΔY_1, ΔX_2 and ΔY_2 are obtained as follows:

$$\Delta X_1 = \Delta Y_2 (\tan\theta_1 + \cot\theta_2) \simeq 1/(S\cos\theta)$$

$$\Delta Y_1 = \tan\theta_1 \simeq \tan\theta$$

$$\Delta X_2 = \cos\theta_1/S_1 \simeq \cos\theta/S$$

$$\Delta Y_2 = -\sin\theta_2/S_2 \simeq -\sin\theta/S \quad ,$$

where $\theta_1 = \tan^{-1}\Delta Y_1$ and $\theta_2 = \cot^{-1}(\Delta X_1/\Delta Y_2 - \tan\theta_1)$ are the actual rotation angles in the first and second pass, respectively, and $S_1 = \cos\theta_1/\Delta X_2$ and

```
procedure resampler(Xstart, Ystart);

--(Xstart,Ystart) is the first resampling position.
-- Resampling parameter registers are initialized as follows:
-- case Mode of
-- SinglePass:   {xPositionInc := ΔX_x; yPositionInc := ΔY_x; xStartInc := ΔX_y; yStartInc := ΔY_y};
-- TwoPass.1st:  {xPositionInc := ΔX_1; yPositionInc := 0;   xStartInc := ΔY_1; yStartInc := 1};
-- TwoPass.2nd:  {xPositionInc := ΔX_2; yPositionInc := 0;   xStartInc := ΔY_2; yStartInc := 1}
-- end

begin
    for Yaddr := 1 to OutHeight do      -- OutHeight: height of the Output Buffer
    begin
        x := Xstart;    y := Ystart;
        for Xaddr := 1 to OutWidth do   -- OutWidth: width of the Output Buffer
        begin
            x_0 := Integer(x);   dX := Fraction(x);
            y_0 := Integer(y);   dY := Fraction(y);  -- dY = 0 for TwoPass
            NewPixel := 0;
            zY := dY;
            for j := y_0 to y_0 + k do   -- if SinglePass then k := 1 else k := 0
            begin
                zX := dX;
                for i := x_0 to x_0 + 1 do
                begin
                    if (xmin <= i) and (i <= xmax) and (ymin <= j) and (j <= ymax) then
                        PartialProduct := Lookup( Lookup( InputBuffer[i,j],zX), zY)
                    else PartialProduct := 0;   -- Lookup( p[i,j], z) = p[i,j]*W(z)
                                                -- Do not accumulate if InputBuffer[i,j] is not valid
                    NewPixel := NewPixel + PartialProduct;
                    zX := Complement( dX)   -- Get 1's complement to represent (1-dX)
                end;
                zY := Complement( dY)
            end;
            OutBuffer[Xaddr,Yaddr] := NewPixel;  -- Output resampled pixel
            x := x + xPositionInc;   -- Get next resampling position
            y := y + yPositionInc
        end;
        Xstart := Xstart + xStartInc;   --Get starting position for next scanline
        Ystart := Ystart + yStartInc
    end;
end.
```

Fig.12.12. Definition of chip operation

$S_2 = -\sin\theta_2/\Delta Y_2$ are two scale factors obtained for the two directions. The resampled grid in this algorithm does not form a perfectly square grid unless $\theta_1 = \theta_2$ and $S_1 = S_2$.

A set of 6-bit/pixel gray images have been synthetically resampled and recorded on film in the gray domain. These pictures are shown in Figs.12.13, 14. The fractional part of the resampled space position indicator was limited to 8 bits of significance and the bilinear interpolation kernel quantization parameter was varied from 1 to 4 bits of significance. The one-bit quantization is nearest neighbor and its effects are noticeable, particularly in Fig.12.14. The difference between 3- and 4-bit quantization effects is mini-

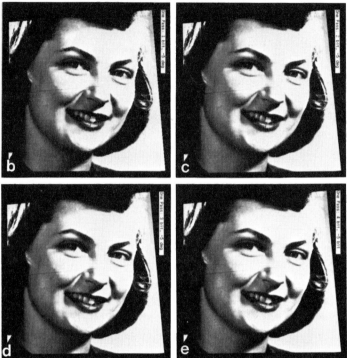

Fig.12.13a-e. Single-pass 5° rotated image using 1, 2, 3 and 4-bit quantized interpolation kernels. a) Scanned original; b) 1-bit quantized kernel; c) 2-bit quantized kernel; d) 3-bit quantized kernel; e) 4-bit quantized kernel

mal, hence for practical systems 3 bits of quantization appear to be sufficient. It should also be noted that generally the gray resampled image is subsequently halftoned to transform tonal modulation to a spatial rendering. This process tends to mask imperfections as well as add its own artifacts.

Fig.12.14a-e. Two-pass 45° rotated test patterns using 1, 2, 3 and 4-bit quantized interpolation kernels. a) Scanned original; b) 1-bit quantized kernel; c) 2-bit quantized kernel; d) 3-bit quantized kernel; e) 4-bit quantized kernel

The performance of the resampling system is determined by the subsystem and disk access performance. The resampling subsystem is estimated to be capable of generating resampled pixels at a rate in excess of one megapixel per second. This component has a minor effect on the overall throughput when compared with the disk access performance. Figure 12.15 is a graph of the disk

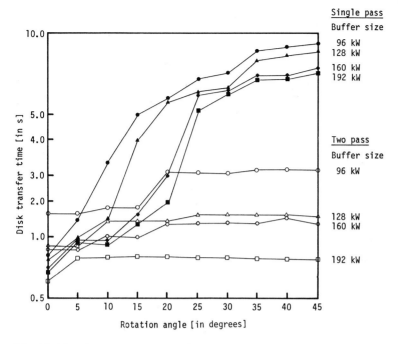

Fig.12.15. Disk transfer performance

transfer time as a function of rotation angle and available buffer memory space for a full page display-oriented operation. The parameters used are as follows:

Image Size: 8" × 10", 80 pixels/inch
(640 scan lines, 800 pixels/scan line)
Scaling Factor: 1
Transfer Rate: 35.4 Mbits/s (formatted)
Inter File Seek Time: 30 ms
Intra File Seek Time: 10 ms
Rotation Latency: 8.33 ms.

It can be seen that the performance of the one-pass system decreases with increasing rotation angles, while that of the two-pass system is "constant." The one-pass system is better at small rotation angles, whereas the two-pass system is better at large angles. In both cases, the performance is between very good and adequate response time for the user performing interactive processing at the CRT display.

12.5 Trade-Offs and Conclusions

The functional difference between the single-pass and two-pass algorithms is that the single-pass can accommodate a nonseparable kernel function, while the two-pass cannot. Images have been synthetically resampled both ways using a separable four-value approximation of the bilinear interpolation function. The resampled images were recorded on film in the gray domain and the results were quite satisfactory, particularly when compared to the results obtained by resampling with the nearest-neighbor kernel. In terms of computational complexity the single-pass algorithm exhibits n^2 growth and the two-pass behaves linearly with n, where n is the number of pixels to be interpolated along a scan line. From the performance point of view, the single-pass has a systematic advantage by virtue of the ability to handle the data only once and compute in parallel, as opposed to a two-stage systematic "pipeline" computation as in the two-pass. The complexity and performance issues are related in that they can be traded off against each other in an implementation sense equalizing the benefits. The two approaches exhibit differing artifacts. The single-pass approach is characterized by a static error in the desired versus actual angle of rotation and scale factor while being able to handle square grids on input and resampled output. Meanwhile, the two-pass approach exhibits unequal static along/across line rotation angle and scale factor errors, resulting in the inability to create a perfectly square resampled grid, as well as occasionally interpolating over a parallelogram rather than square area. It must be emphasized that these static errors are under a fraction of a percent, expected to be quite adequate for commercial applications.

The system performance implications provide the most crucial distinctions. The features here have been gauged in terms of system performance and resource requirements. The major distinctions originate from disk handling. The single-pass system is suitable to small angle rotation, up to about $5°$; for larger angles, the performance rapidly deteriorates. In juxtaposition, the two-pass system is suitable for arbitrary angle rotation as its performance is roughly "constant" with angle of rotation. While the performance generally improves with the addition of main memory, the performance curve shapes remain similar.

There are several disk-related implications. The two-pass operation requires additional storage for the intermediate formatted images; it also is sensitive to disk formatting. Since segments must be easily accessible, the disk sector size should ideally be compatible with this. Extrapolating the performance profiles based on disk handling, it would appear that a lower data rate disk would have a lesser impact on the two-pass performance as the primary factor is head movement, not data rate. Similarly, it would appear that the two pass would benefit most from a fixed head disk.

From the foregoing it can be seen that the technology exists for building commercially feasible post-scan image processing systems. The rapid technological advances in electronics will soon enable even more powerful and sophisticated systems.

References

12.1 Z. Stroll, S. Kang, S. Miller, V. Kadakia: "Post Scan Image Processing System Study", Xerox Internal Report, Accession No. S81-00026 (August 1981)
12.2 R.H. Caron, S.S. Rifman, K.W. Simon: "Application of Advanced Signal Processing Techniques to the Rectification and Registration of Spaceboard Imagery", Technology Transfer Conference Proceedings, University of Houston (September 1974)
12.3 K.W. Simon: "Digital Image Reconstruction and Resampling for Geometric Manipulation", Symposium on Machine Processing of Remotely Sensed Data Proceedings, W. Lafayette, Indiana (June 1975)
12.4 W.K. Pratt: *Digital Image Processing* (Wiley, New York 1978)

Subject Index

Adaptive kernel convolution   16
Algorithmic analysis   140
Alpha network   201
Alpha structure   202,206
Array processor   133
Asymmetric vector reduction   71
Asynchronous   152
Asynchronous timing   136

Backward chaining search   159
Basic systolic cell   19
Bilinear interpolation kernel   224,228
Binary image   176
Binary tree   2
Boundary detection   178

CAD/CAM system   188
Candidate architectures   200
Cardinality   195,205
CCC   206
c-dotted rule   90
Cell borders   179
CELLSCAN   175
Cellular arrays   123,137
Cellular logic processor   175,185
CFG   205,206
CLF recognition   88
Cluster analysis   65
Cluster center   67
Cluster center matrix   68
Cluster center updating process   68,80
Clustering   46

Communication channel utilization   195
Component tree   204
Composition bitmap   211
Computation flow graphs   196,199,200
Computer utilization   205
Computer vision   195
Computing power   204
Computing resource graph   204-206
Concurrency   139
Concurrent systems   107
Condensed frame generator   197
Condenser   197
Context-free grammar   88
Context-free language   87
Convolution   9,214
Convolution with the sinc function   209
Coordinate transformation   213,215
Counting and sizing   178
Counting of residues   180
Covariance matrix inversion   27,28,35
CRG   205,206
Cropping   212
Cube connected cycle network   201
Cube connected cycles   201
Curvature constraint   158
Curve detection   157
Cutoff frequency   176

Data-driven multiprocessor   152
Data flow machines   139
Data transfer   199
Data type of arrays   181

231

Decision-theoretic approach 86
Decoding 210
Decompression 210
Degree of parallelism 199
Design layout 192
Diff series of machines 176,182
"Diff3" 182
Diff3-50 182
Diff4 182
Difference picture 198
Difference picture generation 197
Discrete Fourier transform 37
Discriminant functions 33
Display master 211
Distributed computer system 196
Dotted rule 88
Dynamic programming 46,53,157
Dynamic scene analysis 199,207

E-beam technology 153
Electronic darkroom 211
Electronic master 209
Error-correcting CFL recognition 88
Euclidean distance 46
Exploratory pattern analysis 65

Faddeev algorithms 119
FFT 137
Figure of merit function 158
Fixed kernel convolution 16
Floorplan 187
Full-array automata 175
Full connected network 201
Full resolution gray master 211
Fully connected network 201

Gaussian distribution 33
Givens rotation 144
Global synchronization 142
Golay logic processor 180
Golay parallel pattern transform 180
Golay primitives 183

Graph-theoretic clustering 66
Graphics composition/page makeup 209
Gray-level image processing 176

Halftoning 209,211,212
Hexagonal neighborhood 182
Hierachical clustering 66
Hypercube topology 201
Hypertree 204
Hypertree structure 203
Hypertree topology 202

ICARUS 188
Image correlation 149
Image processing 133
Image processing array 37
Image resampling 209,210,213
Indexed vectors 181
Interchip (interpackage) 40
Interleaved multiple vector reduction 72
Interpolation 213
Interpolation kernel function 214
I/O bandwidth 10,26,32,40
I/O limitation 27
Ionic 111
ISODATA 66

Kernel function 228

Label reassignment process 68,78
Labeled DP generation 197
LANDSAT 209
Language called GLOL 181
Laser programming 153
Least-square error solver 143
Levenshtein distance 46,53
Linear systolic array 9
Link bandwidth 204
Liver tissue 177
Logical convolutions 177
Logical neighborhood 176

Logical processing 176
Logical transforms 176
"Look back" register 178
Lookup structures 114
Lookup tables 183
LSI/PHP 186
L-U decomposition 28

Massively Parallel Processor 137,175
Matrix algorithms 140
Matrix Data Flow Language (MDFL) 142
Matrix multiplication 27,140
Mean vector 35
Memory redundancy feature 183
Mesh connected network 202
Microcoded and vectored 182
Minimum-distance classification 45
Motion analyzer 197,199
Motion detector 197,198
Multichromatic image processing 178
Multicomputer architecture 200
Multicomputer system 195
Multiple subarray processors 182
Multiple thresholding 176
Multiplication Oriented Processor 118
Multitree structure 203

n-cube connections 202
Nearest neighbor 224
Nearest-neighbor array 202
Non-numeric computation 53
Nonparametric cropping 209
Number-theoretic 109

Object extractor 199
One-D array 2
One-dimensional linear array 2
One-pass resampler 215
Optimal curve 157
Orthogonally separable interpolation 214

Parallel computers 138

Parametric cropping 211
Partitioned array 39
Pattern analysis array processor 33
Pattern matching 53,150
Pattern matrix 67
Pattern primitives 86
Pattern vector 86
Performance 207
Performance evaluation 195,206
Permutation matrix 218
Pipelined array processors 138
Pipelined processing 80
Pipelining 141
Preston-Herron processor 183
Processor utilization 195
Programmable systolic chip 20
Pseudoparallelism 195

QR decomposition 144
Quad-tree 2
Quadratic discriminant function 36

RADIUS processor 117
Reconfigurable array 26,30,33
Reconfigurable square array 27
Rectangular array 2
Recursive 139
Recursive algorithm 150
Region classifier 197,199
Region grower 197
Resampling 9
Resampling kernel 209
Residue arithmetic 114
Rotation transformation 213

Scaling transformation 213
Scheduling 196,205
Scheduling algorithm 205
Seismic discrimination 60
Selectable kernels convolution 17
Shape analysis 178
Shuffle-exchange network 2

233

Sidelobes 176
Signal processing 140
SIMD array machine 157,202
Single-pass 210
Skeletonization 178
Skewed storage 76,79
Software metrics 111
Speedup 195,205,206
Splines 48
Squared-error pattern clustering 66, 67,69
String distance 45
Structured language 181
Subfield register 184
Subsystem functional flow diagram 222
Super-computer 136
Symbolic computation 113
Symmetric vector reduction 71
Syntactic pattern recognition 87
Systolic architecture 66
Systolic arrays 9,34,70,118,135,138, 157
Systolic convolution arrays 11
Systolic pattern clustering array 82
Systolic searching tree 1
Systolic system 138

Template matching 38
Texture analysis 178
Three-dimensional architecture 123
Throughput 195
Time-warping distances 46,47,59

Toeplitz 119
Toeplitz System Solver TOPS-28 121
Translation transformation 213
Triangular array 29
Triangularization 145
Turnaround time 195,205,206
Two-dimensional hexagonal array 2
Two-dimensional square array 2
Two-level pipelined architecture 73
Two-level pipelined systolic arrays 22,66
Two-pass method 210
Two-pass resampler 211,215,216

Variable line length 184
Vector operations 70
Vector reduction 71
Very Large Scale Integration (VLSI) 1, 133
Virtual pattern matrix 75
VLSI 195
VLSI array architecture 25

Wafer-scale integrated processor 153
Wafer-scale technology 123
Wave fronts 141
Wave-front arrays 135
Wave-front propagation 148
Wave propagation 142
Weighted Levenshtein distance 47,57
Weiner-Levinson 120

X-tree 1

Very Large Scale Integration (VLSI)
Fundamentals and Applications
Editor: **D. F. Barbe**

2nd corrected and updated edition. 1982. 147 figures. XI, 302 pages
(Springer Series in Electrophysics, Volume 5)
ISBN 3-540-11368-1

Contents: *D. F. Barbe:* Introduction. – *J. L. Prince:* VLSI Device Fundamentals. – *R. K. Watts: Advanced Lithography.* – *P. Losleben:* Computer Aided Design for VLSI. – *R. C. Eden, B. M. Welch:* GaAs Digital Integrated Circuits for Ultra High Speed LSI/VLSI. – *E. E. Swartzlander:* VLSI Architecture. – *B. H. Whalen:* VLSI Applications and Testing. – *D. F. Barbe, E. C. Urban:* VHSIC Technology and Systems. – *R. I. Scace:* VLSI in Other Countries. – Addenda. – Subject Index.

Charge-Coupled Devices
Editor: **D. F. Barbe**

1980. 133 figures, 7 tables. XI, 180 pages
(Topics in Applied Physics, Volume 38)
ISBN 3-540-09832-1

Contents: *D. F. Barbe:* Introduction. – *G. J. Michon, H. K. Burke:* CID Image Sensing. – *W. D. Baker:* Intrinsic Focal Plane Arrays. – *D. K. Schroder:* Extrinsic Silicon Focal Plane Arrays. – *D. F. Barbe, W. D. Baker, K. L. Davis:* Signal Processing with Charge-Coupled Devices. – *J. M. Killiany:* Radiation Effects in Silicon Charge – Coupled Devices.

Pattern Formation by Dynamic Systems and Pattern Recognition

Proceedings of the International Symposium on Synergetics at Schloß Elmau, Bavaria, April 30–May 5, 1979

Editor: **H. Haken**
1979. 156 figures, 16 tables. VIII, 305 pages
(Springer Series in Synergetics, Volume 5)
ISBN 3-540-09770-8

Contents: Introduction. – Temporal Patterns: Laser Oscillations and Other Quantum-Optical Effects. – Pattern Formation in Fluids. – Turbulence and Chaos. – Pattern Recognition and Pattern Formation in Biology. – Pattern Recognition and Associations. – Pattern Formation in Ecology, Sociology and History. – General Approaches. – Index of Contributors.

Springer-Verlag
Berlin
Heidelberg
New York
Tokyo

Digital Pattern Recognition
Editor: **K.S.Fu**
With contributions by numerous experts
2nd corrected and updated edition. 1980.
59 figures, 7 tables. XI, 234 pages
(Communication and Cybernetics, Volume 10)
ISBN 3-540-10207-8

Contents: Introduction. – Topics in Statistical Pattern Recognition. – Clustering Analysis. – Syntactic (Linguistic) Pattern Recognition. – Picture Recognition. – Speech Recognition and Understanding. – Recent Developments in Digital Pattern Recognition. – Subject Index.

Syntactic Pattern Recognition, Applications
Editor: **K.S.Fu**
With contributions by numerous experts
1977. 135 figures, 19 tables. XI, 270 pages
(Communication and Cybernetics, Volume 14)
ISBN 3-540-07841-X

Contents: Introduction to Syntactic Pattern Recognition. – Peak Recognition in Waveforms. – Electrocardiogram Interpretation Using a Stochastic Finite State Model. – Syntactic Recognition of Speech Patterns. – Chinese Character Recognition. – Shape Discrimination. – Two-Dimensional Mathematical Notation. – Fingerprint Classification. – Modeling of Earth Resources Satellite Data. – Industrial Objects and Machine Parts Recognition.

Light Scattering in Solids I
Introductory Concepts
Editor: **M. Cardona**
2nd corrected and updated edition. 1983.
111 figures. XV, 363 pages
(Topics in Applied Physics, Volume 8)
ISBN 3-540-11913-2

Contents: *M. Cardona:* Introduction. – *A. Pinczuk, E. Burstein:* Fundamentals of Inelastic Light Scattering in Semiconductors and Insulators. – *R. M. Martin, L. M. Falicov:* Resonant Raman Scattering. – *M. V. Klein:* Electronic Raman Scattering. – *M. H. Brodsky:* Raman Scattering in Amorphous Semiconductors. – *A. S. Pine:* Brillouin Scattering in Semiconductors. – *Y.-R. Shen:* Stimulated Raman Scattering. – Overview. – Additional References with Titles. – Subject Index. – Contents of Light Scattering in Solids II, III and IV.

Springer-Verlag
Berlin
Heidelberg
New York
Tokyo